DISASTER MAKERS

AF291503

DISASTER MAKERS

Tackling Unmanaged Growth for Sustainable Futures

Terry Gibson

BLOOMSBURY ACADEMIC
LONDON • NEW YORK • OXFORD • NEW DELHI • SYDNEY

BLOOMSBURY ACADEMIC
Bloomsbury Publishing Plc
50 Bedford Square, London, WC1B 3DP, UK
1385 Broadway, New York, NY 10018, USA
29 Earlsfort Terrace, Dublin 2, Ireland

BLOOMSBURY, BLOOMSBURY ACADEMIC and the Diana logo are
trademarks of Bloomsbury Publishing Plc

First published in Great Britain 2025

Copyright © Terry Gibson, 2025

Terry Gibson has asserted his right under the Copyright, Designs and Patents Act,
1988, to be identified as Author of this work.

For legal purposes the Acknowledgements on p. xiii constitute an
extension of this copyright page.

Cover design by Chris Bromley
Cover image © Melinda Nagy via Adobe Stock

All rights reserved. No part of this publication may be reproduced or transmitted
in any form or by any means, electronic or mechanical, including photocopying,
recording, or any information storage or retrieval system, without prior permission in
writing from the publishers.

Bloomsbury Publishing Plc does not have any control over, or responsibility for, any
third-party websites referred to or in this book. All internet addresses given in this
book were correct at the time of going to press. The author and publisher regret any
inconvenience caused if addresses have changed or sites have ceased to exist, but
can accept no responsibility for any such changes.

A catalogue record for this book is available from the British Library.

Library of Congress Cataloging-in-Publication Data
Names: Gibson, Terry (International development researcher), author.
Title: Disaster makers : tackling unmanaged
growth for sustainable futures / Terry Gibson.
Description: London ; New York : Bloomsbury Academic, 2025. |
Includes bibliographical references and index.
Identifiers: LCCN 2024026783 (print) | LCCN 2024026784 (ebook) |
ISBN 9781350430471 (hardback) | ISBN 9781350430488 (paperback) |
ISBN 9781350430495 (epub) | ISBN 9781350430501 (ebook)
Subjects: LCSH: Disasters–Economic aspects. |
Economic development–Environmental aspects. |
Risk management. | Hazard mitigation.
Classification: LCC HC79.D45 G53 2025 (print) | LCC HC79.D45 (ebook) |
DDC 363.34–dcundefined
LC record available at https://lccn.loc.gov/2024026783
LC ebook record available at https://lccn.loc.gov/2024026784

ISBN: HB: 978-1-3504-3047-1
PB: 978-1-3504-3048-8
ePDF: 978-1-3504-3050-1
eBook: 978-1-3504-3049-5

Typeset by Newgen KnowledgeWorks Pvt. Ltd., Chennai, India
Printed and bound in Great Britain

To find out more about our authors and books visit www.bloomsbury.com
and sign up for our newsletters.

CONTENTS

Part 1
UNNATURAL DISASTERS

Part 2
UNMANAGED GROWTH

Part 3
SUSTAINABLE FUTURES

Chapter 13

ILLUSTRATIONS

Figures

Map

Tables

ABBREVIATIONS

AFL	Action at the Frontline (Project of GNDR)
ASJ	La Asociación para una Sociedad más Justa' – Association for a More Just Society
CO_2	Carbon dioxide
CSO	Civil Society Organization
DRR	Disaster Risk Reduction
ESG	Environmental, Social and Governance monitoring
FAO	Food and Agriculture Organization (of the United Nations)
FORIN	Forensic Investigations of Disasters Project
GAR	Global Assessment Report on Disaster Risk Reduction (published by UNDRR)
GDP	Gross Domestic Product
GND	Green New Deal
GNDR	Global Network for Disaster Reduction (full title Global Network of Civil Society Organizations for Disaster Reduction)
GPMB	Global Preparedness Monitoring Board (Joint project of WHO and WB)
IPCC	Intergovernmental Panel on Climate Change
LMIC	Lower Middle-Income Country
MNC	Multinational Corporation
PPE	Personal Protective Equipment
SDGs	Sustainable Development Goals
SFDRR	Sendai Framework for Disaster Risk Reduction (UNDRR framework)
UK	United Kingdom
UN	United Nations
UNDRR	United Nations office for Disaster Risk Reduction (Previously UNISDR)
UNESCO	United Nations Educational, Scientific and Cultural Organization
UNFCCC	United Nations Framework Convention on Climate Change
VFL	Views from the Frontline (Project of GNDR)
WB	World Bank
WHO	World Health Organization

ACKNOWLEDGEMENTS

The journey of this book spans several decades and along the way a number of people have helped me develop my understanding of the worlds of development, disaster studies and writing.

Stephen Rand (Tearfund); Lucy Figeroa, Jo Ann van Engen and Peter Clark (Honduras); Patricia Norrish and Pete Mann, who navigated me through my doctorate; John Norton (Nepal, Vietnam and other projects); Marcus Oxley and the secretariat and members of GNDR; Development and Disaster Studies experts including Bina Desai, J. C. Gaillard, Ilan Kelman, Allan Lavell, Mark Pelling and particularly Ben Wisner; those who took the time to read it and provide valuable suggestions including the anonymous reviewer, Ben, Bina, Tim Grew and Andrew Hardwick; my editor on this and my previous book, Nick Wolterman; the editorial and production teams at Bloomsbury; most of all, my wife Olwen.

I'm grateful for all their help and support. I haven't listened to everything they said and the shortcomings and deficiencies of this work are entirely my own.

PREFACE: A PERSONAL DISASTER JOURNEY

The clue's in the title. This is a book about disasters. At the heart of it is the idea that disasters are not natural. Disasters large and small result from human choices and the scale, cost and impact of them are increasing. Most visibly in the media, the future consequences of climate change are increasingly being reported. The wider effects of unmanaged growth are increasing the risk of disasters for all of us. How can these trends be tackled? How can we secure sustainable futures? This book explores ways to make the world a safer and more prosperous place for all of us … and for our children and our children's children.

For many, disasters are an ever-present reality. Some, like me, are more fortunate and are rarely exposed to their effects. The 2020 Covid-19 pandemic, however, touched everyone. 'Doomscrolling' was a new buzzword of the time – the action of constantly scrolling through and reading depressing news. Living through that experience and becoming absorbed in what seemed a constant stream of bad news, people became sensitized to further disasters. Headlines post-pandemic provided plenty of doom to scroll through. Some suggest that populations were afflicted by a 'collective trauma' resulting from the events they'd endured.[1] Surveys showed that right across Europe and North America three-quarters of the population were anxious about the effects of global warming.[2] The future, which had seemed so full of the promise of greater prosperity even for those in less developed countries (at least according to the global institutions that monitor these things), seemed to be heading on a downward track towards potential catastrophe (a word derived from the Greek meaning 'downward turn'). What will the future look like for the next generation? Is there anything we can do about it? Is it possible to change course and shape a safer, happier, prosperous world for all its people?

The ideas and insights I have drawn on come primarily from the world of disaster studies, but this book is my own take on them, so I need to explain why, as someone who never set out to be a 'disasterologist', I've come to write it, particularly since living where I live (the UK) and doing what I do, I have, generally, not been much affected by disasters.

My early career at the beginning of the 1990s was in communications, producing training and marketing videos for large companies, along with occasional TV programmes. I had, however, a niggling feeling that I wanted to see more. It might have originated from a church youth group I went to, which was concerned with faith and by implication with morals and ethics. We sometimes watched filmstrips (VHS was about to burst onto the world but hadn't quite made it yet) from a newly formed aid agency, 'Tearfund'. Unusually, their work, though church-related, was very different from the old-style missionary projects typical of that time. They talked about justice, economics, the environment, poverty and how those things needed to be tackled to make the world a fairer, more equitable place.

That niggling feeling persisted, maybe fuelled by those filmstrips. I am not a natural adventurer. In fact, I hate travel, or at least everything up to the moment I finally board a plane. But nevertheless I wanted to go, see, do something about it. I was put in contact with Stephen Rand at Tearfund. He was similarly niggled and had given up a teaching job to organize education (things like those filmstrips) at the agency. It was thanks to him that I eventually became a regular film-maker for Tearfund (VHS video was now a reality) alongside my portfolio of corporate customers. The great thing about working for Tearfund was that education, understanding the challenges of development and justice, was really important to both Stephen and the agency.

Let's get back to *disasters*. Stephen wanted to produce a package which helped people to understand them. The source material was a book on disaster management published by Tearfund[3] and its centrepiece was the 'Crunch' diagram. I had no idea then how big a part this simple diagram would play, later, in my work and thinking. One of the book's authors was Ian Davis. He'd been introduced to me by Stephen. He was a quiet-spoken architect whose route into disasters had been through disaster housing. It was he who, in his book on that subject, created the 'crunch diagram'.[4] Read the rest of this book to find out about the concept in more detail, but at its simplest the diagram shows that disasters are created at the crunch point of *hazards* and *vulnerability*. *When It Comes to the Crunch* (1993)[5] was the pack we created, with a big bold crunch diagram on the cover. I wrote the study guide to accompany it and for a few months had a fleeting familiarity with the subject.

What I didn't realize until many years later was that Ian Davis was working with his colleagues Piers Blaikie, Terry Cannon and Ben Wisner on writing what would become a core work in the emerging world of disaster studies – *At Risk: Natural Hazards, People's Vulnerability and*

Disasters.[6] Published in 1994 it built on the 'Crunch' diagram, renaming it the *Pressure and Release Model* and expanded it with chapters of theory and rich case studies. I was oblivious to its existence.

I embarked on my next project and the crunch was quickly forgotten. In our business we had to become instant experts on any subject we made films about … and then move on. But something stuck. One word: 'vulnerability'. Those made *vulnerable* by poverty and lack of resources were far more impacted by *hazards* than others. I came to see the localities I visited while film-making for Tearfund through new eyes. Over a decade or so centred on the 1990s I visited, filmed in and stayed in many villages, towns and cities across Africa, Asia and Latin America. Over that period, I only twice came close to major disasters: in the devastated bombed-out city of Kabul in 1993 and in reconstruction after Hurricane Mitch in Honduras, in 2002. However, the message of *When It Comes to the Crunch* was writ large in many of the localities I visited: in rural villages in East Africa, remote settlements in North Peru, urban slums in Phnom Penh and New Delhi, marginalized communities on Bangladeshi islands, garbage villages in Cairo. In these and many other places I recognized a different kind of disaster: not those intense, dramatic events that hit the headlines but much smaller, regular events: what I would later call 'Everyday Disasters'. These were repeated, grinding shocks and stresses resulting from vulnerability to small-scale hazards – seasonal flooding, failed crops, poor or non-existent healthcare, malnutrition, lack of access to clean water, the scourge of AIDS, the less publicized scourge of malaria and, overarching it all, grinding poverty – small everyday disasters that never hit the headlines.

By the time I worked on a project about the aftermath of Hurricane Mitch in 2002 – a suite of films covering different aspects of that disaster and the learning from it[7] – I realized there was a whole vocabulary and conceptual framework around disasters of which, apart from my early encounter with the 'crunch', I was completely unaware. The in-country team in Honduras introduced me to some of it, which made me realize I didn't know what I didn't know; so I set out to pursue a part-time master's degree in International Development, which led on to a part-time doctorate. My study reassured me that some of my sense-making of my many visits did make sense and through a chain of coincidences it boomeranged me back into the world of disasters.

I needed a case study as a focus for my doctorate. It was about ways of learning and sharing knowledge through networks. Two possibilities had foundered. In frustration I happened to look at Tearfund's jobs page and found that a new network they were hosting, more or less independent of them, was looking for a project manager majoring on communications.

'The Global Network for Disaster Reduction' (GNDR) was newly formed with only one staff member, its chief executive Marcus Oxley, who had been seconded from Tearfund. I became the second member of its tiny secretariat (2008). Alongside gathering data for my doctorate as a 'participant observer' I embarked on the most exciting and stimulating period of my professional life. We networked together hundreds of small civil society organizations (CSOs), dotted across the globe, typically with very small but highly motivated staff teams. They were united in their concern for working on disaster risk reduction (DRR) among local communities. They collaborated with us to gather and share data and insights which we used with them to campaign and lobby the United Nations system for disaster reduction. Our flagship 'Views from the Frontline' projects and reports[8] – based on consulting thousands of communities and local organizations in many disaster-affected countries – gathered a huge amount of data and evidence on the need to address disaster reduction at the local level … and to address everyday disasters.

I'd previously felt inarticulate and clueless in the world of disaster reduction but now I learnt day by day from our network members, people on the ground who were getting their hands dirty doing DRR at the sharp end. I was also learning from my colleague and boss, Marcus, who'd headed up the function at Tearfund before taking on this network and was phenomenally experienced, a clear-thinker and very driven. And it was after I started working for the network that the 'crunch' finally resurfaced. One of the network's advisers – an expert, activist and co-author of 'At Risk', was Ben Wisner, who became a colleague, mentor and friend.

My early intuition has stayed with me. Disasters at all scales need to be addressed but the impact of everyday disasters resulting from exposure to small-scale regular hazards – to which local populations are made vulnerable to through lack of resources – is at the heart of the story. These persistent impacts grind down peoples' lives and increase the impact of major disasters too.

Alongside that enduring idea I also learnt through my many encounters that the 'hazards' story is more complex than those early diagrams suggested. Alongside the physical hazards which we think of as typically implicated in disasters, human activity is both *magnifying* the impact of such hazards and *creating* entirely new ones. New risk is being created. Disasters are being made. This seemed to me the most important trail to follow and it's the one I've pursued in this book. It begged a huge question. Who, or what, is making disasters? Activist and poet Ai Weiwei was confronted with the awful reality of the Sichuan earthquake in China, in which sixty-nine thousand people died – nineteen thousand of them schoolchildren, crushed when badly built schools collapsed. He was moved to say:

The disaster makers always go free and the innocent are punished.[9]

Who, or what, are the disaster makers? Is disaster risk created by malignant individuals or organizations, as the term suggests? Or are wider and historic forces at play? Is the creation of ever-increasing risk the result of a wider system? How can this growing disaster risk be addressed? Spoiler alert! My answer is there in the book's subtitle: *Tackling Unmanaged Growth for Sustainable Futures*. But there are two ways in which this answer is nuanced. Firstly, the alternative to unmanaged growth doesn't have to be political or economic revolution. I will argue that there are better alternatives within the system as it is, or ought to be. Secondly, I don't think the starting point is through concentrating on 'fixes' for unsustainable aspects of the system. It lies, I will argue, in articulating visions for a better future, and then setting out with all the creativity, ingenuity and innovation of humanity at its best to work out how to get there. We'll meet again at the end of the book. See you there!

INTRODUCTION

Many think of 'natural disasters' as the unavoidable consequence of violent natural events. This book's title suggests that, as is often said, 'no disaster is natural' and that in some way disasters don't just happen but are *made*. This book is driven by the claim – made by those studying disasters – that disaster losses, despite our best efforts, are increasing and that this trend is driven by creation and magnification of disaster risk. This claim is deeply disturbing. We can see how the last two centuries have been ones of phenomenal industrial, technological, economic and social progress and many graphs can be drawn of positive trends associated with this transformation. But if the findings of disaster studies are correct, then there are other negative trends in play which may increasingly undermine this progress. It's important, therefore, to assess these claims and identify the 'disaster makers' driving them. The journey of this book is in three parts.

Part 1: Unnatural disasters

Disasters aren't natural; disasters often aren't even large in scale and they don't affect everyone equally – these and related ideas have been explored and developed by those working in disaster studies, often deeply involved at the sharp end and concerned to move beyond simple response to disaster events and discover root causes. We trace the development of their thinking to build an understanding of the true nature of disasters, focused at the nexus of 'hazards' and 'vulnerability' and magnified in many ways by human actions and choices, as a basis for finding out why losses and disaster risk, despite efforts to manage them nationally and internationally through UN frameworks, are increasing.

Part 2: Unmanaged growth

Accumulating risk of disasters, large and small, is the result of human activity. To understand why this is the case we need to take a historical journey spanning the 'Anthropocene' period, during which human actions are thought to be changing the face of our planet. We explore three phases of the Anthropocene: from the start of the Industrial Revolution in the 1760s, from the start of the consumer revolution in 1908 and from the start of a 'great acceleration' in trends of human-caused changes from around 1950. Our focus turns to the state of world cities – home to an increasing proportion of the world's population – 68 per cent by 2050. The goal is to discover the relationship between human activity and growing disaster risk, despite the insights and goals of global frameworks designed to manage it. This is most visible not only in the widely recognized impact of greenhouse gas production on our climate but in many other less publicized ways too.

Part 3: Sustainable futures

Three key actors are identified in the Anthropocene journey: *industry* (by which we mean the whole industrial complex of innovation, production, branding, marketing, finance and growth), the *public* (not only as participants in the consumer cycle but as citizens, the essential units of humanity) and *government*, in its role of governance of this cycle of industrial, economic and social growth and change at local, national, international and global levels. The cycle driven by these three actors has yielded dramatic technological, economic and social transformations, resulting in better livelihoods, health and longevity for many. Despite these achievements, half the world's population still live in poverty,[1] and creation of risk through overdevelopment, emissions, pollution and social segregation is increasing in growing world cities. Unmanaged and unsustainable development threatens to engulf us globally. Who are the disaster makers? We conclude that while individuals and organizations acting out of self-interest play a part, we are all implicated, though the poor often have little choice. Where, in this constantly growing cycle, is it possible to break in, take control and steer our world away from increasing disaster risk and towards a sustainable and prosperous future? The attitudes and roles of citizens in driving change based on positive future visions are key to this.

Part 1

UNNATURAL DISASTERS

Chapter 1

ENCOUNTERING DISASTERS

The fragility of normality

Picture the moment a disaster strikes. Life has continued day after day, then, suddenly, everything changes. A moment of shock, disbelief, disorientation. Filming and working with people in many localities round the world I have heard survivor stories many times. Such events are immediate, shocking and dramatic, coming without warning. Lydia's story was particularly striking.

> When it was seven at night my husband told me, 'Lydia, get up because I hear a horrible noise coming and it must be the current.' 'No, you're silly', I told him. 'Come, lie down.' I set him down in bed, got him plugs to put in his ears and said to him 'come over here, stop bothering me'. And I covered his ears too. He said to me 'You are the crazy one. Take these things out of my ears, because a current is coming.' When he said that all of a sudden I felt the side of my waist and it was wet. The water was up to the side of the bed.
> (Lydia Mercedes Alvaredo, *Survivor of Los Zanjones*)[1]

We were filming interviews with survivors of this disaster – three years after the event – in Posoltega village, Nicaragua, in Central America, the narrow strip of land joining North and South America. The village lies on a fertile rural plain north of the country's capital and consists of a number of settlements farming cattle, maize and cotton. In 1998 Hurricane Mitch, the most powerful storm in the region for decades, made landfall in Honduras. It was Thursday the 29th October.[2] In its capital, Tegucigalpa, seven of the ten bridges spanning the main river Choluteca which ran through the city were destroyed, along with many buildings. Flooding and landslides caused huge destruction right across the country, and over eleven thousand people are thought to have died. The storm moved slowly west into Nicaragua and the event Lydia experienced hit global headlines.

The noise Lydia tried to shut out was the oncoming roar of the huge mudslide which had shot down the flank of the nearby Casitas volcano, across the Nicaraguan plain to the sea, flooding, crushing and burying everything in its path. The crater in the summit had filled rapidly from storm rains and when part of it sheared away the mudslide hurtled down the hillside at almost one hundred kilometres an hour in a three-metre-high wave of mud, water and volcanic debris. Two of the area's settlements, El Porvenir and Rolando Rodriguez, were in its direct path and were completely buried under the mud. Rescuers could only identify their location by GPS. About twenty-five hundred people lost their lives, mostly in those two settlements.[3] Lydia was lucky, relatively speaking, that she and her community were not on the mudslide's direct path, but the whole community was shattered by this sudden disaster which left bodies strewn across the plain and land and homes buried beneath the mud. Headline news led to a massive response and even President Clinton visited, but the huge aid response couldn't bring back peoples' homes or loved ones.[4]

I stood on the dried mudslide, vegetation gradually covering it now, and heard survivor stories as we filmed with a team trying to capture what happened and how to learn from it. Even standing there and looking across the plain to the deep scar in the volcano, the events of that night seemed unimaginable. I could sense the shock of the event and well understood how Lydia would have rejected the awful possibility until her bedroom started flooding. Faced with the fragility of normality we try to cling on and shut out other possibilities.

Nearly two decades later I was guided around coastal towns on Japan's north-east coast which were still being rebuilt after the devastating Great East Japan earthquake and tsunami of 2011.[5] A massive, one-in-a-thousand-year event, the submarine earthquake lifted the ocean floor by several metres and pushed a massive tsunami onshore in less than thirty minutes, crashing over sea walls and funnelling the waters into harbours, bays and valleys. The enormous, fast-moving wave, three to seven metres in height, pushed inland to flood Sendai, the region's major city. Many who had trusted the sea walls or felt safe on higher land were inundated. The Fukushima power plant was engulfed, the reactors shut down and leaking radioactive material, a nuclear catastrophe narrowly averted.[6] When we visited the area four years later, in 2015, we found many homes and factories had been abandoned to decay, the area around Fukushima still a no-go zone. In other areas large-scale rebuilding was underway. One of the rail routes in the region only reopened during our stay. In one coastal port we witnessed the ground across the entire valley being raised by shifting huge quantities of material from the nearby hills. A constant stream of earthmovers rumbled up and down the tracks, like a string of ants building their nest. Here and everywhere were stories of loss. One of our guides pointed out a walkway sloping up from a coastal plain to higher ground. He still found it difficult to describe how they tried to help older people up that slope as the tsunami roared towards them, but some just could not make it. Over nineteen thousand people lost their lives as a result of the disaster.

My work took me to another earthquake-prone region, Kathmandu, Nepal, in 2014, a year before the major Gorka earthquake. I took a 'disaster walk' with colleagues from a specialist earthquake technology agency, who showed me a monument to the previous 1934 earthquake and explained the likely consequences of the next big one. I returned in 2017[7] and found that their predictions of loss, damage and death in the overcrowded and badly built capital were sadly correct. Buildings crumbled, streets were blocked, people fended for themselves as emergency services struggled to reach them and international aid also struggled to reach the country through its one, damaged, international airport. Nearly nine thousand people died.[8]

It's uncomfortable visiting these events as an observer, remaining dispassionate, partly because filming and documenting require a kind of detached objectivity. Nevertheless, survivors' stories of that moment of shock when everything suddenly changes strike a deep emotional chord. I guess that just as Lydia refused to believe the evidence of her ears at the actual moment of disaster, all of us cling onto fragile normality until overwhelming evidence makes us realize that everything has suddenly changed, we've got to take our earplugs out, feel the floodwaters around us and face the reality of the sudden, shocking disaster event.

Global disaster: The pandemic

I said that I'd avoided experiencing a disaster until recently. That all changed in 2020. As a secure inhabitant of a North European country I was accustomed to think of disasters as remote events to observe and study. I remember only too well the sudden shock of deserted streets, daily briefings and isolation of the Covid-19 pandemic sweeping across the world, no-one immune from its effects. For once I, and many others, experienced unbelievable scenes streaming from the world's media – lockdowns, silent cities and makeshift morgues in China, across Asia, the disease's landfall in Europe. Eerie scenes in Italy as regional borders were closed and refrigerated trucks became morgues. The government seemed to be in denial in the UK as the pandemic approached, assuring us that we were well prepared, and first rejecting and finally accepting the advice of scientists for a complete shutdown of our country.[9] Suddenly, certainties evaporated. Would we die? How would we eat? What would happen next? Would we ever again lead normal lives? Initial denial gave way to continual doomscrolling through statistics and briefings, increasingly feeling part of a sci-fi disaster movie. If I had failed to empathize sufficiently with disaster survivors I met previously, I certainly did now. The impact on most peoples' lives (including mine) was far smaller than that of some of the disaster events I've referred to, though globally nearly 7 million deaths were reported by late 2023[10] and economic and social consequences continue to reverberate. Attuned to disaster there seemed a cascading stream of them in the following years – supply chain problems and shortages, war in the Ukraine, energy crises, food shortages, price hikes, extreme weather, the 2023 Turkey/Syria earthquake, calamitous war in Sudan and brutal battles in Israel and the Gaza Strip.

Acts of god?

Headline disasters are characterized by suddenness, intensity and unexpectedness. This is nothing new. An event from Roman times generated striking, stark images of the immediacy and power of a disaster. The eruption of Mount Vesuvius in 79 CE engulfed the city of Pompeii under lava, ash and poisonous gases in a matter of minutes.[11] The city lay buried and forgotten for many centuries, until archaeologists dug into the debris and discovered voids in the ash which, when cast in plaster, revealed the city's inhabitants frozen in time at the moment of death. That sense of sudden shock is captured in the poses of figures cast into plaster of Paris, buried in lava and volcanic ash by the eruption. A group of thirteen figures in an orchard seem to be caught trying to flee. A dog writhes on its back. A single figure is sat with its hands covering its eyes. The casts capture a moment of absolute horror.[12]

These are devastating events defined by their enormous, intensive scale and effects, their unexpectedness, their unpredictability. Much closer to the present day, at 4.17 am on 6 February 2023, as most people slept, a devastating earthquake, the largest there since 1939, struck Turkey and Syria. About 1.5 million people were made homeless and sixty thousand died. Turkey's president referred to the earthquake as a case of force majeure – irresistible force[13] – and terms such as 'Act of God' and 'Black Swan Event' were used. In fact the very word 'disaster' comes from the Latin *Dis Astrum* meaning 'Bad Star' – something malevolent falling on us from above.[14] But are these events really unanticipated?

Unpredictable disasters? Risk is being created

As I've walked through the sites of many disasters and met their survivors something doesn't ring true. Are disasters really natural, unexpected, unpredictable, unavoidable? There may be those who want us to see them like that, malignant 'bad stars', with inevitable impacts. But looking more closely at almost any seemingly 'natural' disaster reveals a different story. A colleague of mine went to help in the response to the 2010 Haiti earthquake which flattened much of its capital, Port-au-Prince. He told me he could crumble the concrete columns of collapsed buildings in his hands, sand substituted for cement so builders could save money and boost profits.[15] Weak steel beams had ruptured and folded, yet the same type of beams was being used in new constructions. The earthquake itself was severe, but people died because badly built structures crushed them. Similar stories emerged from the Turkey/ Syria earthquake. Very quickly after the event people started to ask why some buildings collapsed while neighbouring structures withstood the earthquake. It emerged that though Turkey's Izmit earthquake of 1999 had led to new earthquake-safe construction codes, over half of Turkey's buildings evaded these codes, and worse, five years before the latest earthquake, construction companies were offered amnesties that authorized their illegal, earthquake-unsafe buildings for a fee, leading the country's Chamber of Civil Engineers to say, 'It will mean transforming our cities, notably Istanbul, into graveyards.'[16]

Frozen images of the moment of death in disaster-struck Pompeii suggest instant, unanticipated disaster, but actually even in that case there were warnings. Several earthquakes had rocked the city in the previous two decades, there was evidence that buildings were still being repaired at the time of the catastrophe and further evidence that many of the wealthier inhabitants had left, spurred by signs of volcanic activity in the days before the disaster. Much of this was recorded by Pliny the younger who escaped the disaster but watched his own father sail back into the city under the shadow of the smoking volcano, to perish.[17]

Many disasters can, it seems, be anticipated: even the pandemic which engulfed us all in 2020. In the UK the possibility of a flu-type pandemic stood at the top of the national risk register,[18] alongside repeated warnings and reports from the World Health Organization[19] and the Global Preparedness Monitoring Board.[20] There had already been four smaller regional pandemics in the first two decades of the twenty-first century.[21] In the UK, major simulation exercises had taken place in the preceding years, from exercise 'Winter Willow' in 2007 to exercise 'Cygnus' in 2016, though the latter and an even more important one dealing specifically with a coronavirus threat ('Exercise Alice') were kept secret until the middle of the Covid-19 pandemic.[22] Despite the findings of these exercises highlighting the need for coordination, protection of those in care and maintenance of adequate stockpiles of personal protective equipment (PPE), the coordinating body for response to such a disaster was wound down and scrapped six months before the pandemic.[23] Stockpiles of PPE were reduced by one-third in the three years before the event.[24]

Examining the backgrounds to other disasters reveals an emerging pattern, which like the backstory to the pandemic suggests that disasters are rarely unanticipated.

The shocking destruction of the communities of el Porvenir and Rolando Rodriguez by the Casitas mudslide during Hurricane Mitch might seem impossible to predict, but actually right across Nicaragua and Honduras the level of destruction from floods and mudslides was magnified by massive deforestation. At that time between 100,000 and 120,000 hectares of forest disappeared every year,[25] and in the period between 1990 and 2010 the UN FAO estimates Nicaragua lost 31 per cent of its forest cover.[26] In the Posoltega area, in common with many others in the country, forest was stripped out to make way for farming. The loss of forest cover removed a natural brake and barrier to floods and mudslides and reduced the stability and water-carrying capacity of the deforested areas. People looking for land and livelihoods colonized increasingly precarious situations. The two communities buried under the mudslide on the slopes of the volcano had only been established a few decades previously, pushing outwards from the village to find farmland.[27] The human impact of the hazard – the *disaster* – resulted from people being pushed towards land made even less safe by massive deforestation.

The human impact of the Great East earthquake and tsunami of 2011 in Japan was also a result of choices about where to live and work, of trust in sea walls built to protect against 'expected' tsunamis but not against a one-in-a-thousand-years event. Those walls were not only engulfed by the tidal wave but then trapped the floodwaters behind them. The protective walls at the Fukushima nuclear plant were only built to the height of the previous tsunami, not to reflect future trends and predictions.[28]

The Nepalese Gorka earthquake struck a densely overpopulated city, the result of continual inward migration from impoverished rural regions magnified by civil war. Buildings that according to codes established after the previous 1936 earthquake should have been only four stories high had been subsequently extended to six, seven, eight stories to increase rents for absentee landlords who didn't bear the cost of the increased risk. These buildings, too, had been built in many cases without adhering to earthquake-safe building codes.[29]

Even the ancient Pompeii disaster has thoroughly modern echoes. The area exposed to a future eruption of Vesuvius is, today, heavily populated. The way risk zones are mapped by the civil defence authorities has allowed construction of the region's largest hospital *inside* the risk zone for major eruption, because the method of mapping zones is based on political negotiations about boundaries rather than volcanological estimates.[30] In all these cases human choices lie behind disaster impacts – creating unnatural disasters.

Defining disasters: Intense … and everyday

Let's take a moment to define disasters, and in doing so make our first contact with the industry concerned with them. While the Concise Oxford English Dictionary defines disaster as 'a sudden accident or a natural catastrophe that causes great damage or loss of life'[31] the United Nations Office for Disaster Reduction (UNDRR) had a more nuanced and shifting definition. In 2004 it defined them as 'A serious disruption of the functioning of a community or society causing widespread human, material, economic or environmental losses which exceed the ability of the affected community or society to cope using its own resources.'[32] The nuances I highlight are the suggestion in the UN definition that the event might strike a *community or society* – so the scale of the disaster might be large or small – and the reference to 'serious disruption' rather than 'catastrophe', again allowing for different scales of disasters. OK, you may feel these are fine distinctions, but remember that the United Nations is very careful with their choice of words and what I see here is the very slight opening of a door to a broader definition in terms of scale: 'community or society' and of impact: 'serious disruption'. This becomes clearer if we wind forward to 2023 and look at the current UNDRR definition:

> A serious disruption of the functioning of a community or a society at any scale due to hazardous events interacting with conditions of exposure, vulnerability and capacity, leading to one or more of the following: human, material, economic and environmental losses and impacts.
>
> Annotations: The effect of the disaster can be immediate and localized, but is often widespread and could last for a long period of time. The effect may test or exceed the capacity of a community or society to cope using its own resources, and therefore may require assistance from external sources, which could include neighbouring jurisdictions, or those at the national or international levels.[33]

This definition is different again, in that 'the effect may test or exceed the capacity of a community or society', whereas the previous definition asserted that a disaster *would* exceed the ability of the affected community or society to cope using its own resources. But why does this matter? Isn't this just an exercise in semantics? Actually, I think it signifies a pivotal shift in UNDRR's understanding of disasters. In our initial encounters with disasters the scale has been large, the events dramatic. This latter definition allows for disasters which are not only localized but may not even require external resources to address them. These are disasters of a very different scale, and of great interest to myself and the people I work with as we have seen, particularly through the data we gathered through the GNDR 'Views from the Frontline' (VFL) projects, that many of the disaster events that affect peoples' lives are *not* large-scale intensive disasters but small-scale *everyday disasters*. For example, a colleague in Nigeria took me out into a rural, forested area in Akwa Ibom state, along a rough unmetalled road, to an effective dead end where a timber-built bridge across a river had collapsed. The slumped structure was passable on foot but no vehicles could get across and the villages beyond were largely starved of supplies. He and his small organization had been collaborating with local government to try and raise funds to rebuild the bridge, but the villagers had been waiting years for this to happen. Though hitting no headlines this – for the local villages affected by it – was a disaster.[34]

Everyday disasters: Accumulating risk

The idea that all disasters are large-scale sudden events was being undermined by my encounters. For example, rural Tanzania with its many relatively isolated bush villages ought to have been safe from the challenges of AIDS in other parts of Africa, but along the 'Truckers Route' – the main trade route through Tanzania to landlocked Zambia – the disease was being spread by infected lorry drivers using sex workers who had come to the roadside villages to work, and then returned to rural communities carrying infection, which because basic health centres could not detect or treat it then spread within the villages.[35]

In West African Benin's commercial capital, Cotonou, a densely populated suburb, Cocotomey, suffered from flooding for seven months in each year after the seasonal rains. Nearly a metre of flood water closed the schools, clinic and the market area, affecting education, health, commerce and access to supplies. This regular everyday disaster had been allowed to persist by the local government, who locals said only took notice at election time, promising action until they were elected, and then forgetting again.[36]

In North-east Africa, on Eritrea's border with Sudan, Amar, along with her fellow-villagers, faced the latest drought and crop failure. She had once owned a flock of five hundred goats. When drought came, she had to sell goats to buy food, so her flock diminished. Every time this happened, everyone else was trying to sell goats too so the price was driven down and she got a poor price for the animals. Now, as I interviewed her, she was sat with her last goat, expecting to be dependent on aid from now on.[37] These and many other 'everyday disasters', lying under the radar of agencies and the media, knock back peoples' lives regularly and insidiously.

The persistent nature of regular shocks and losses, disrupting life and work, from these everyday disasters seemed to have a greater impact on those exposed to them than the threat of a major disaster. Outside the coastal city of Limbe in the West African state of Cameroon I stood on the solidified lava rock at the edge of the flow from the last (2000) eruption of Mount Cameroon. The city was at risk from the regular eruptions (three in the last century) – both from the possibility of the lava and ash reaching the city and from the extreme weather events an eruption could trigger. Nevertheless, when local people in the city were consulted in a GNDR 'Views from the Frontline' survey about the threats that concerned them, the main concerns of poorer people living in the low-lying area of the town – where housing was affordable – were flooding, followed by access to basic services, access to water, landslips (a consequence of seasonal rains), poverty and unemployment.[38]

Some of those working on that survey wondered whether these threats even counted as disasters, but when we presented some of our work to a global reinsurance company and I mentioned our sense that these 'everyday disasters' were dominant, one of their team said that while they focused on major disasters they found, year-on-year, what they called 'attritional risk' – resulting from smaller scale regular events – accumulated and outweighed headline disasters in terms of the costs they had to insure.[39] What's more, as we will see when we return to the story of Hurricane Mitch in the next chapter, it is the accumulation of everyday disasters which magnifies the risk of an intense disaster. The persistent erosion of people's lives as they deal with the repeated impact of small-scale disasters impoverishes them, damages their property and possessions and affects their livelihoods. If an intense disaster strikes, their capacity to recover afterwards is weakened. Everyday and intense disasters are connected.

Local heroes: Local action and risk reduction

The news so far has not been good. For another, more positive perspective, let's join a small meeting at a conference centre in the hills overlooking Kathmandu for a few days in October 2017, to share experiences of *resisting* disasters. The participants had converged from different directions: from the Pacific Islands, the Philippines, Vietnam, Indonesia, India, Nepal, Pakistan and Cameroon. They all had stories to share of working to reduce peoples' risk exposure. Though I've called them, tongue in cheek, 'local heroes', they had no magic wand, no silver bullet. They shared in common dogged determination to work together with local people, identifying the causes of everyday disasters in their localities and fighting against them. For example, in New Delhi a CSO got involved in a locality in the east of the city, an area of shanty dwellings where many of the poor lived, working in the barter economy, 'unauthorized' and disregarded by the authorities. Pollution, seasonal flooding, disease and crime were the grinding everyday disasters that wore them down.

The CSO drew people together into the 'East Delhi Citizens Watch Forum', itself a lengthy and difficult task as people weren't used to collaboration. They gathered evidence of the challenges people faced, mobilizing young people and even created an app for them to report what they found. They mounted a campaign targeting the authorities and the media and gradually persuaded local government to take action on challenges such as improving drainage and policing. Their key methods were to involve community members, give them voice, gather evidence and challenge local authorities. Other participants in the meeting described promoting typhoon-safe housing in coastal Vietnam, creating safe employment on Philippine islands and establishing sustainable crops – in the face of climate-change-driven drought – in rural Indonesia. These and other stories showed how these 'local heroes' could work with local populations to drive down risk.[40]

This is the world of 'disaster risk reduction' (DRR) – getting smart and taking action before disasters to reduce their impact. It's also the world that we saw UNDRR starting to recognize, one of small-scale 'everyday disasters' which are sometimes addressed by local action rather than outside intervention. There is a caveat to these good news stories however. The evidence of UNDRR and others is that *new* risk is being created much faster than it is possible to manage and reduce it. While there are huge numbers of stories of *risk reduc*tion like those recounted above, there's a much bigger systemic problem of *risk creation*.[41] For example, our colleagues from Multan in Central Pakistan, attending the Kathmandu workshop, described how their attempts at reducing risk from flooding from the Chenab river were undermined by wealthy landowners diverting floodwaters away from their land towards poorer communities, and also explained that flooding was amplified further by deforestation upstream in the northern hills, and further still by changing weather patterns driven by global warming.[42] Local heroes collaborate with communities to reduce risk where possible, but beyond those localities other actions and other actors, outside their control, are constantly generating new risk.

Who pays? Not those who create the risk

If disasters are not just 'bad stars' falling from above, unpredictable and uncontrollable, if they're not just the results of dramatic hazards colliding with humanity, if they are somehow the consequence of human action and if they seem to seek out the poorer and more vulnerable, then who, or what, is driving this collision of humanity with hazard and leading to disasters which harm or kill those least able to protect themselves?

An earthquake struck the Sichuan region of China in May 2008, destroying many thousands of buildings. Of the sixty-nine thousand people who died, some estimates are that as many as nineteen thousand students died as over seven thousand school buildings in the region, clearly not built to be earthquake-proof, collapsed. Shock turned to anger as the collapse of these buildings and the scale of loss of schoolchildren gradually trickled into the media.[43]

Chinese poet and dissident Ai Weiwei wrote an impassioned critique of the governmental failures which resulted in those children being crushed in their thousands, saying 'The disaster makers always go free and the innocent are punished.'[44]

It is those others, the innocent poor, the vulnerable, who bear most of the consequences. Put simply, disasters don't happen, they're made. In many cases those in positions of power generate increasing risk, as was the case in choices about planning laws, standards of construction by government and attempts to evade them by builders in the Sichuan earthquake. They offload the risk they profit from onto those unable to protect themselves from it.

How can we unravel all of this? If unnatural disasters are the consequence of human choices, then how can we tackle the creation of risk? How can we build on the work of local heroes, managing and reducing risk? How do we make the juggernaut change course? What do we need to change in our society, our governance, our economy, in the way we lead our lives to achieve this? We need to take a step back and find out what people working in the field of disaster studies have discovered over the decades they've been working in this field.

Chapter 2

THE RISK EQUATION

No disaster is natural … introducing Mitch

The sudden, unexpected shock that Lydia experienced as floods roared towards her house in Posoltega, Nicaragua, as Hurricane Mitch struck, on that October evening in 1998, is one of the striking characteristics of many large disasters. Their immediacy and scale galvanize the attention of response agencies, media and the public. People respond with compassion, generosity … and also with fascination. Images from such disasters fill our screens for days on end. This compulsive interest spawned the whole disaster movie genre: Films such as *San Andreas, Deep Impact, Contagion, A Perfect Storm* and *Towering Inferno* are all reliable box-office staples. That's one way we try and make sense of disasters.

As I flew out of New York towards Honduras during work on the Hurricane Mitch project in late 2001, a pall of smoke rose below me. My journey had been delayed by several days as flights had been grounded after the 9/11 attacks on the World Trade Centre twin towers. A week after the event smoke was still rising from the site as we flew over it. This real-life event dwarfed disaster movies. Footage of the plane strikes and collapsing towers played back again and again like a loop tape. It was sudden, unexpected, devastating, dramatic. What's more, unlike many disasters observed afar by people in the rich, developed world it was in their backyard, throwing into doubt their security as well as their understanding of global geopolitics, ideologies and cultures. The newfound ability of smartphones to capture and broadcast events as they happened from multiple perspectives, fresh and unedited, contributed to the corporate shock, devastation, anger … and fascination.

A few years later the film *Contagion* (2009) eerily anticipated the pandemic we were all to live through – to the extent that the then UK

Health Secretary Matt Hancock claimed he turned to it for inspiration in tackling that crisis.[1] There's a whole social imaginary of disasters, sometimes blurring real and fictional, as we seek to make sense of them.

Undoubtedly, 9/11 was anything but natural. By contrast, the Hurricane Mitch disaster I was documenting with colleagues was more easily understood and assimilated in the media and public imagination as a 'natural disaster', happening 'over there' in another world, one where 'things like that happen'. But even disasters such as Mitch, distant though they often seemed to people in developed countries, were gradually drawing increasing attention. Not just in cinemas and on the TV news, but in the corridors of governments and international agencies. Institutional concern was driven by hard facts. The prevalence, impact and cost of disasters was increasing. Between 1980 and 2000 the number of 'natural catastrophes' recorded worldwide rose by over 50 per cent, according to global re-insurer Munich Re.[2] The World Bank estimated that losses doubled from US$50 to 100 billion over that period.[3] Concern coalesced into the launch of the 'Decade of Natural Disaster Reduction' in 1990 by the UN, intending to drive reduction in their impact and cost.[4] While governments and institutions were waking up to the increasing human and financial cost of disasters, people closer to the actual events were starting to question the use of the word 'natural' with its implication that these were 'bad star' events, outside human control.

In this and the following chapters Hurricane Mitch provides a case study to help chart the development of thinking about the nature and causes of disasters. Mitch is a decades-old disaster, but precisely because we can look both at what *preceded* it and what happened in the years *afterwards* it's a good case to consider. While media and public attention to disasters may last days or weeks, disaster impacts last years and decades. More recent disasters don't give us the same opportunity to look at the whole pre- and post-disaster timeline. I have a personal interest too, as I'd worked in Honduras on a number of occasions between 1990 and 2015, giving me a slight sense of the life of the country and 'earthing' me through these brief encounters, helping me, and hopefully the reader, to avoid over-abstracting and theorizing. Personal experience helps keep people's places and lives at the forefront of one's mind, just as we all have very personal memories of the recent pandemic disaster. Disaster studies isn't just dispassionate, observational science. It is, or ought to be, driven by concern and solidarity, with an emotional as well as a rational dimension.

Map 1 Map of Honduras (adapted to show locations discussed). 2014, Courtesy: EU Echo. CC4.0.

The story of Honduras, before, during and after the storm, provides a basis for charting the developing understanding of disasters, their causes and effects.

Before the storm: Accumulating risk

Take a light aircraft flight over the protected Honduran rainforest landscape of Mosquitia – the ground moving so slowly below that you seem to be almost hovering– and you'll see the problem. The problem for the inhabitants of this country is that much of Honduras, like Mosquitia, is formed of mountains, valleys and rivers, with few plains or plateaus. Mosquitia, in the north-east corner of the country, is still hidden under forest, as the whole country once was. Apart from muddy rivers weaving among the contours, every square metre is covered with dense rainforest vegetation. Even seemingly sheer escarpments have trees impossibly clambering over them. Honduras and its neighbours, Costa Rica, Guatemala and El Salvador, jostle shoulder-to-shoulder on the thin Central American isthmus connecting North to South America, sharing this challenging geography. They share a dramatic climate, too. Intensely wet rainy seasons, hot dry seasons and the annual storm season, often battering the coastlines and countryside with hurricane-force winds.

As Honduras's population grew, pressure for access to farmland displaced many. Of the agricultural land that is workable much was appropriated by big agro-business. A 1993 estimate was that the Honduran government and two banana companies – Chiquita and Dole – owned approximately 60 per cent of the cultivable land in the country.[5] Only 1 per cent of farmers held 25 per cent of the land in large farm estates ('latifundios'),[6] while 70 per cent of farmers held just 10 per cent of the land in small landholdings ('minifundios'). Many small farmers were pushed onto poorer marginal land in forests and on hill and mountain slopes and then faced the destruction of their homes when storms struck. In 1990, eight years before Hurricane Mitch, these forces were already shaping the lives of the people and their environment. Many were fighting for places to live and survive.

Nueva Suyapa, for example, is an informal township established on the edge of Honduras's capital city, Tegucigalpa. People displaced by an earlier storm, Hurricane Fifi in 1974, found this land, a former rubbish dump, and settled on it. From here, high on the edge of the city, they could see its sprawling centre below them, framed in a bowl in the mountains. At the city's heart were slick hotels and arcades where those who'd made it into the prosperous elite came to shop and dine, before retreating to safe substantial properties in the hills outside the city.[7] Life is very different in Nueva Suyapa – or 'La Infiernito' (*little hell*) – as one policeman described it.[8] Ignored by the city authorities for decades, the township was built around dirt roads on steep, unstable slopes. For many years it lacked proper water supplies, waste disposal, healthcare or schooling. Families fought to build lives here and find work. Pressures the young faced, lacking education and employment options, turned many to crime. Drug dealing and gun crime were endemic. Countering this, community associations had grown with small initiatives, for example, providing internet access so that family members working abroad could keep in touch, which led to boosted remittances back home. People knew what they needed, it was just very hard in this poor and neglected community to get it.[9]

Away from the city, life was very different. On the north coast, away from the cities, the Garifuna ethnic group lived in coastal villages such as Santa Rosa de Aguan. This small township of a few hundred people, established nearly a century ago, was based on fishing, small-scale arable farming and cattle ranching, set behind coconut-tree-fringed sand dunes and an idyllic-looking Caribbean-style beach. Though remote from Trujillo, the nearest city, it did have basic services. It was a calm and peaceful place from day to day, but the township faced both environmental and social challenges. Sandwiched between sand dunes protecting it from the sea, a lagoon and the Aguan river, it was exposed to the risk of storms and floods battering it from all sides. As the population grew, houses and streets ate into the protective dunes, exposing them to the sea. At the same time farmers moving into the area grabbed rights to inland areas, squeezing the small community between the two.[10]

Away from the coast in the centre of the country the Olancho region, once an entirely forested area, had been gradually opened up, useable land stripped of forest for farming. Local small-scale farmers, 'Campesinos', mixed cattle ranching with corn, bean and sorghum production. However, like most areas of the country the best farming land had been taken over by big agro-business, backed in many cases by national government schemes and foreign investment. Pressures on small farmers meant they cut away more forest to create farmland. Some turned to other businesses, particularly carpentry, using yet more of the local timber to create doors, windows and furniture. Many left the area altogether, either looking for opportunities in other regions of the country or migrating to Mexico and the United States, hoping to find work and send remittances back home.[11]

City, coast and inland areas were all under pressure, shaping the people and the land. Just one corner of the country remained under some protection from these forces: Mosquitia – also known as 'The Mosquito Coast' – in the north-east corner of the country, the one remaining substantial area of rainforest. It was populated largely by indigenous groups – the Tawaca, Pech, Garifuna and Misquito. From Puerto Lempira, Mosquitia's capital, itself accessed in the 1990s only by light aircraft or from the sea, reaching the interior of the region was only possible by river. Krausirpe village on the river Patuca had somehow attracted the attention of development workers and ethnographers despite or maybe because of its inaccessibility (it was beyond the reach of Coca-Cola, a practical definition of remoteness). When I joined this steady stream of 'development tourists' and took the dugout canoe journey to the village it seemed settled and secure, under the protection of its UNESCO 'Biosphere' status. A village elder, Isidro, painted a picture of an easy-going local life of fishing, small-scale hunting, arable farming of plantains, bananas, rice, manioc and beans for food, and cacao as a cash crop.

Even this remote area, however, was facing the pressures of population and business. The dense rainforest was being attacked by loggers coming to extract the tropical hardwoods, ranchers clearing land for cattle farming, and large- and small-scale farmers, many from Olancho, claiming rights to parcels of land. Krausirpe and other settlements in the region were vulnerable, as in many cases they didn't have official titles to their land, exposing them to others snatching their territory. Forestry workers and a locally based CSO painted a bleak picture of the future as these pressures increased and local populations struggled to defend and adapt their lifestyle.[12]

Nearly a decade before Hurricane Mitch, people in the cities, towns and villages of Honduras were facing social, economic and environmental pressures, competition for land and livelihoods, a squeeze on their lifestyles as migrants and big business grabbed the land, pushing them to carve out new farmland, stripping the country of its forests and displacing many to the cities or to emigrate in search of work – all this quite apart from the possibility of any major external shock.

How disaster studies emerged in response

It was the complexity of the relationship between everyday and major disasters, and the underlying pressures which drove both, which led to the emergence of *Disaster Studies*. The story of the accumulation of risk is told thousands of times in different ways, particularly across Latin America, Africa and Asia. The problems might be different but the underlying issue of persistent social, economic and environmental pressures squeezing ordinary people is a common theme. Recognizing this, in the 1970s and 1980s a mixed bunch of development workers, aid workers, geographers, social scientists, urban planners and architects who were all looking at disasters from different perspectives gradually got together. They realized that those pressures combined with external hazards and it was the two together that resulted in disasters.

In retrospect those seemingly random disciplines make a lot of sense. Disasters are complicated, they're not just about meteorology or volcanology, they're about people, places, lots of other '-ologies'. I'm struck by the ways that geographers and architects in particular have to be polymaths to do their jobs. Geography is not just maps and architecture is not just buildings. Geographers engage with society, with politics, with the environment. Architects blend design with technology and weave them into social and cultural contexts. What's more, both need to be earthed in experience, in practical encounters. These are people who are prepared to get their hands dirty.

One of the early pioneers of disaster studies was someone who spent much more time in the field than in the lecture theatre. Fred Cuny is described as a larger-than-life character. His worldwide encounters with humanitarian catastrophes made him realize that disasters and development were intertwined. He was one of the first to shift focus from supposedly 'natural' events to the *vulnerability* of people to hazards, often forced by poverty, migration and pressure to live in cheap but high-risk areas in unplanned, cramped urban areas.[13]

The same obsession with field study drove Henry Quarantelli,[14] a quieter, more thoughtful character, based at the 'Disaster Research Centre' he'd created in the States. He was passionate about figuring out what people did when disasters struck. He challenged the popular idea that disaster victims were powerless and just panicked. He found they were far from powerless and often acted very rationally and collaboratively in response and recovery. Cuny and Quarantelli were joined by other early thinkers in the emerging disaster studies discipline who were also forming connections between hazards and social and economic contexts. A study of *Classic DRR publications*[15] gives as examples *Qui se nourrit de la Famine?* (1975), *The Environment as Hazard* (1978), *Interpretations of Calamity from the Viewpoint of Human Ecology* (1983), *Natural Disasters. Acts of God or Acts of Man?* (1986) and *Disasters, Development and Environment* (1994). The direction of thinking was clearly towards the magnification of disasters by human action and this informed the work of activists working in the field.

Based in the Philippines, Zenaida Willison ('Zen' to her many colleagues) worked at the 'Citizens Disaster Response Centre', where their field studies showed clearly that peoples' 'vulnerability' resulted to a large extent from poverty and the political situation and that these factors accounted for the huge impact of the typhoons and other natural hazards which struck the country.[16] At the same time Mihir Bhatt, who trained as an urban planner, had just founded the 'All India Disaster Mitigation Institute' which was drawing similar conclusions about the effects of disasters on people in his own country.[17] In Latin America Andrew Maskrey and Allan Lavell had been particularly struck by the ways disasters resulted from social forces creating vulnerability, exposing people to hazards. They launched a pioneering regional network – 'La Red' – as a focus for action on disaster reduction in the region.[18]

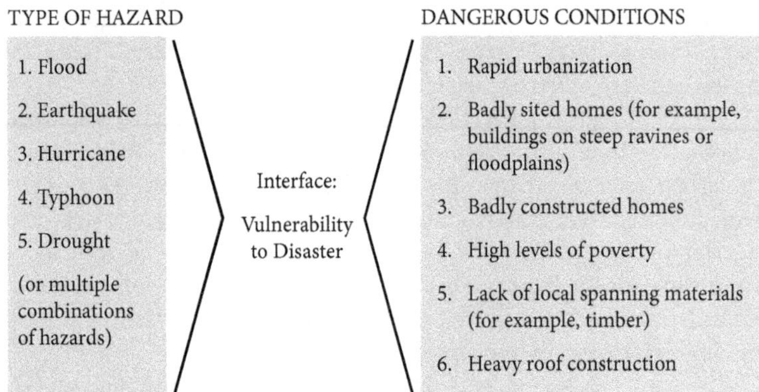

Figure 1 Hazards and vulnerability.
Source: Adapted from Davis 1978, p. 3.

Phil O'Keefe, Ken Westgate and Ben Wisner got together to distil this thinking down into a paper – 'Taking the Naturalness out of Natural Disasters'. It was a neat title to capture the heart of what they were all saying.[19]

Architect and disaster housing expert Ian Davis was deeply influenced by the message of that paper, leading him to write 'Shelter after Disaster',[20] with its simple 'crunch' diagram showing on one side *dangerous conditions*, or *hazards* – which had been the focus of attention in responding to disasters – and on the other side the factors which led to populations being *vulnerable* to hazards. The two forces converged at the centre of the diagram to create the *crunch* – a disaster. It showed how the naturalness was taken out of natural disasters (see Figure 1). You could even say, tongue-in-cheek, that it represented *the discovery of vulnerability*.

The discovery of vulnerability – The crunch model

This developing body of researchers felt that the technocratic approach of the United Nations and many agencies, focusing on 'natural disasters' and response to them, missed the point that it was the vulnerability of populations to various hazards that resulted in disasters. One illustration of this came from a paper Paul Susman had written with O'Keefe and Wisner, taking the example of Hurricane Fifi – the 1974 storm which pushed people onto the land that became Nueva Suyapa. They compared that storm with a similarly severe event in the same year – Hurricane Tracy in Darwin, Australia. While eight thousand died as a result of the Honduran storm, only forty-nine did as a result of Tracy. They saw the contrast as resulting from the different degrees of *vulnerability* of the two societies.[21] But this thinking didn't penetrate the UN's corridors. It was about to launch the 'Decade of Natural Disaster Reduction', keeping the focus on natural hazards and response.[22]

Many years later one of the authors of the 'Taking the Naturalness out of Natural Disasters' paper, Ken Westgate, commented that it took the UNDRR until 2021, more than forty years after that paper was published, to finally acknowledge that disasters are not natural.[23] A press release finally reported: 'This week sees the launch of a new phase in the #NoNaturalDisasters Campaign which seeks to make use of the term "natural disasters" redundant in the workplace' (Mami Mizutori, head of UNDRR, 2021).[24]

This new framing of the nature of disasters was the start of a long journey, not only to develop thinking and translate it into action but to shift the mindsets of people and institutions away from the idea that disasters were unavoidable 'natural' events.

The risk equation

What was it about the economy, politics and people of Honduras, for example, that might be creating the conditions for an oncoming disaster? Building on the work that had been done by these emerging disaster studies researchers in the 1980s, Ben Wisner and Ian Davis had got together with two other disaster studies activists, Piers Blaikie and Terry Cannon, to write what would become a major contribution to understanding the relationship between hazards, risk and disasters: the book *At Risk: Natural Hazards, People's Vulnerability and Disasters*.[25] The four authors traded chapters over nearly five years before publishing their work in 1994. Its publication coalesced much earlier thinking, turning the focus away from 'natural disasters' to spotlight the cocktail of hazards and vulnerability which created the conditions for disasters. At the book's heart was a development of Davis's earlier 'crunch' model (see Figure 2 below). The 'crunch' was as before – the conjunction of *hazards* and *vulnerability*. What was elaborated in this 'pressure and release' model was the sequence of events which developed vulnerability, what the authors called the 'Progression of Vulnerability'. They described this progression leading to *Unsafe Conditions*. Lying directly behind them were *Dynamic Pressures* – such as rapid urbanization, debt repayment schedules and deforestation. Lying in turn behind dynamic pressures were *Root Causes* – systemic problems such as limited access to power, structures and resources, along with the effects of ideologies such as political and economic systems. At the heart of the model was the 'crunch' – the risk of disaster, distilled into a very simple equation: Risk = Hazard × Vulnerability: $R = H \times V$. The model and equation were to form the foundations for the development of disaster studies over the following decades, informing, for example, the understanding of what lay behind the Hurricane Mitch disaster in Honduras.

Risk and real-world vulnerability

The people I've met in my travels and the lives they lead are about much more than an equation. Close-up is different. For example, in a rural village in Tanzania I asked a farmer, Matayo, what his daily life was like and he talked of the weekday dawn-time treks to the fields to manage his maize crops. But he talked also of weekends when he and his family spent time together, met friends, ate and talked. He talked of his family, of how he had met his wife, of the changes in the village where they lived. Of their daughter, who had left, like many, for the city but surprisingly had returned. This is not an equation, these are people. The authors of *At Risk* knew this very well. They recognized that the pressure and release model was static. It didn't tell the story of how life for disaster-affected people unfolded. So they complemented it with the 'access model'.[26] The diagram representing it is quite complex but the underlying message is simple. Sustaining livelihoods for those afflicted by potential disasters, both large and small, involves living through dynamic cycles of events which, if virtuous, may enable people to sustain and build their livelihoods or, if vicious, may persistently erode them:

> Access involves the ability of an individual, family, group, class or community to use resources which are directly required to secure a livelihood in normal, pre-disaster times, and their ability to adapt to new and threatening situations. Access to such resources is always based on social and economic relations, including the social relations of production, gender, ethnicity, status and age, meaning that rights and obligations are not distributed equally among all people. (From *At Risk*, p. 94)[27]

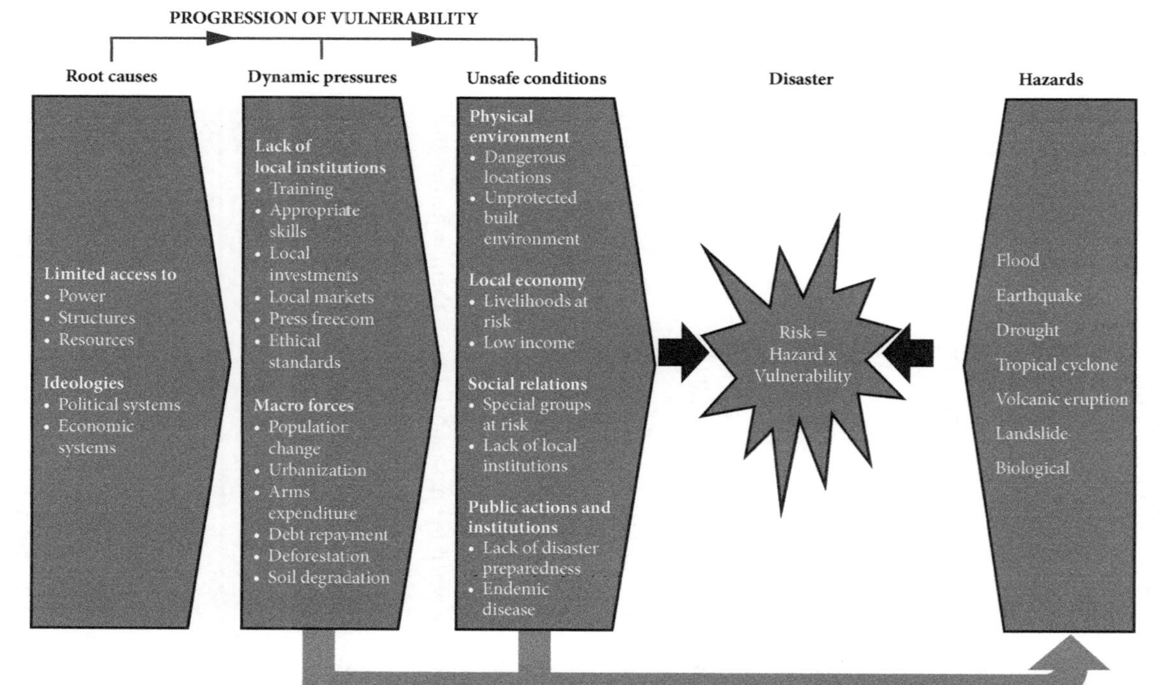

Figure 2 Disaster pressure and release model.
Source: Adapted and simplified from *At Risk*, Blaikie et al. 1994 and *The Routledge Handbook of Hazards and Disaster Risk Reduction* Ch. 3. Wisner et al. 2012.

Matayo and his family had carved out a practical life for themselves, though they worried about droughts and limited access to seed, fertilizer and pesticides. Others in the village were not so prosperous. Lacking land, they worked smallholdings which were insufficient to sustain them. They struggled to find work to supplement their crops and were even more vulnerable to drought. They lacked *access* to the resources to maintain their lives and also to resist and cope with shocks and stresses. 'Access' consists not only of personal assets but also of access to – in this example – agricultural inputs, decent work and also healthcare and education, all of which were only available at a very basic level in the village.

The authors emphasized the dynamic, cyclical nature of their access model, as disasters – in this example the everyday disaster of drought – might not lead to loss of life or total loss of livelihoods but could strip away the few assets villagers held. They might need to sell crops, cattle or land to compensate for their losses and that would impoverish them further, making them more vulnerable to the next shock, driving a vicious circle.

I've previously mentioned the connection between intense, large-scale disasters and smaller, recurrent everyday disasters and this is emphasized in the access model. Moving beyond the static $R = H \times V$ equation, the regular shocks and stresses vulnerable populations experience, whether in the village I visited or in an urban slum, are the everyday disasters which weaken people, taking away their capacity to resist and cope when the next shock strikes them, whether small scale or massive. The experience of vulnerability is harsh and heart-breaking.

Chapter 3

VULNERABILITY

Vulnerability in Honduras: 'I am destroying the land'

The Pressure and Release model depicts a sequence of events leading from root causes to dynamic pressures and then to unsafe conditions – a 'progression of vulnerability'. The access model brings it to life through the cycles of everyday experience which either sustain or erode peoples' livelihoods. In Honduras we have already seen these pressures faced in the coastal village of Santa Rosa de Aguan, in the central Olancho region, in the urban shanty area of Nueva Suyapa and in the remote but increasingly exposed Mosquitia village of Krausirpe. A powerful account of the 'progression of vulnerability' in Honduras was published by Susan Stonich in her book *I Am Destroying the Land!*[1] whose title immediately raises the question *Why*? Why would anyone want to destroy the land? The answer is striking, captured in a quote from a poor Honduran farmer:

> I can only expect destruction for my family because I am provoking it with my own hands. ... This is what happens when the peasant doesn't receive help from the government and the banks. ... He looks for the obvious way out which is to farm the mountain slopes and cut down the mountain vegetation. Otherwise how are we going to survive? We're not in a financial position to say, 'Here I am! ... I would like a loan to plant so many hectares!' I put in my request but the banks don't want to give me credit because I cannot guarantee the loan. I know what I am doing ... as a person I know ... I am destroying the land! (Southern Honduran peasant, 1990)[2]

The perversity of the situation faced by Honduran farmers illustrates powerfully the whole progression of vulnerability. The crunch they face is a combination of hazards – particularly the frequent storms that strike the region – and the progression of vulnerability to those hazards. The farmer knows he is destroying the land through deforestation, increased erosion, the danger of landslides and degraded soil quality. Lying behind those unsafe conditions are the *dynamic pressures* he describes:

Dynamic pressures

- Good quality farmland too costly for poor farmers to purchase from their own resources
- Finance not available to buy good quality land and farm it properly
- Pressure to move to the mountain slopes and cut down mountain vegetation because no other land is available.

Lying behind those dynamic pressures are *root causes*:

Root causes

- Concentration of capital in the hands of wealthy landowners
- Exclusion of poor from financial services

Similar chains of events could be seen in the Honduran localities I described. In the coastal village of Santa Rosa de Aguan the growing population, squeezed by land-grabs on the inland side and with a growing population, were increasingly over-farming the land they had and encroaching further into the seaward protective dunes as they built more homes. They were unavoidably degrading and deforesting land and damaging the dunes just to survive, driven by a progression of vulnerability: the dynamic pressures of limited useable land, growing population and limited resources.[3] In the Olancho region local farmers faced the same challenges as the Southern Honduran farmer quoted above. They were destroying the land as a result of the dynamic pressures they faced.[4] Not only that but they and other larger agro-business concerns were also doing the same in the adjoining Mosquitia rainforest region, logging and opening up cattle ranches, degrading and deforesting the environment there.[5] Others, such as many of the inhabitants of Nueva Suyapa, had been pushed off the land altogether and migrated to the cities where they faced the dynamic pressures of urbanization, limited infrastructure, poor services, lack of education and unemployment driving increasing crime.[6] Figure 3 shows how

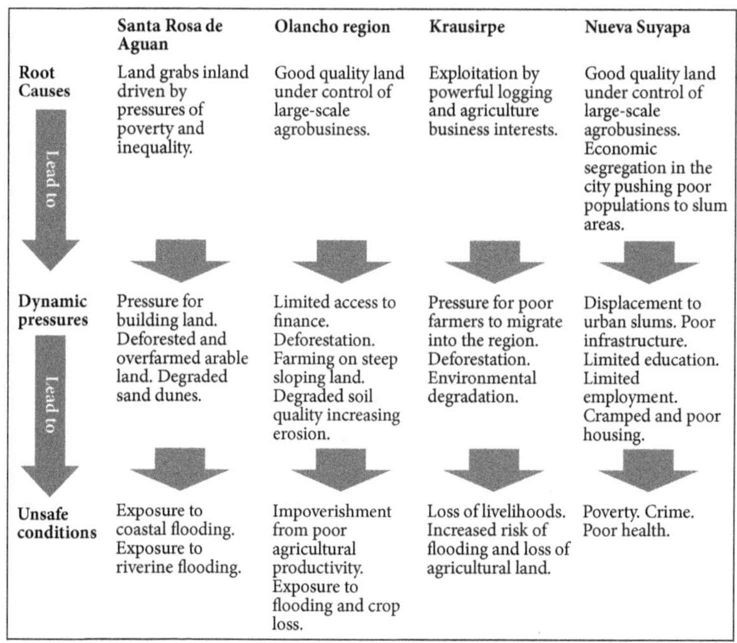

	Santa Rosa de Aguan	Olancho region	Krausirpe	Nueva Suyapa
Root Causes	Land grabs inland driven by pressures of poverty and inequality.	Good quality land under control of large-scale agrobusiness.	Exploitation by powerful logging and agriculture business interests.	Good quality land under control of large-scale agrobusiness. Economic segregation in the city pushing poor populations to slum areas.
Dynamic pressures	Pressure for building land. Deforested and overfarmed arable land. Degraded sand dunes.	Limited access to finance. Deforestation. Farming on steep sloping land. Degraded soil quality increasing erosion.	Pressure for poor farmers to migrate into the region. Deforestation. Environmental degradation.	Displacement to urban slums. Poor infrastructure. Limited education. Limited employment. Cramped and poor housing.
Unsafe conditions	Exposure to coastal flooding. Exposure to riverine flooding.	Impoverishment from poor agricultural productivity. Exposure to flooding and crop loss.	Loss of livelihoods. Increased risk of flooding and loss of agricultural land.	Poverty. Crime. Poor health.

Figure 3 Unsafe conditions and dynamic pressures in four localities in Honduras.
Source: Author.

unsafe conditions and dynamic pressures affected people in all those locations.

The roots of risk

The Pressure and Release model unpacked the forces of vulnerability – *dynamic pressures* queuing up behind *unsafe conditions* and *root causes* in turn bearing down on both – and it also went further, to shift the focus onto *risk*. The risk equation: Risk = Hazard × Vulnerability[7] does two things:

Firstly it brings the key word 'Risk' into the heart of the process. In doing so it moves thinking from the early focus on *hazards* as being the cause of disasters, and further still from the discovery of *vulnerability*, to argue that these two elements in turn configure the *risk* of disasters. While a population as a whole may be equally exposed to a hazard, different sectors may be more or less vulnerable to it, so that the resultant risk may be different for distinct groups in the overall population. The drivers of vulnerability could be seen as *creating* risk for those who experience it. Risk and later risk creation would gradually become central to thinking on the nature of disasters, particularly considering the interaction of social and environmental factors in creating disaster risk.

Secondly, the equation brings a suggestion of scientific rigour to the growing body of understanding on disasters. Science? Was disaster studies now becoming a science? Science is often thought of as having predictive power. Its theories and equations should be capable of suggesting what effects will follow from causes. This is easy enough in the physical sciences where laws predict very precisely the results of chemical and physical interactions. We rely on them for our technological and manufacturing processes. But as far as disasters are concerned, we've already seen that the disaster 'crunch' is the consequence of the combination of hazards and peoples' vulnerability, as the equation suggests. That takes us into the complex worlds of sociology, geography, psychology, economics and politics – all coming into play when we consider peoples' vulnerability. These are fields where we no longer dispassionately observe simple sequences of cause and effect. Unlike a physical scientist studying a phenomenon in a laboratory, we are ourselves part of the social processes we claim to observe. We bring our own cultural and ideological biases to our observations. Not only that, but these processes are no longer simple sequences of cause and effect, but complex cocktails of multiple causes and effects. Can disaster studies really handle this complexity and regard itself as a science?

Some of those working in the field believed so and gradually developed a more 'forensic' approach. In 2008 a consortium of science councils founded the 'Integrated Research on Disaster Risk' programme with the aim of facilitating research on disasters, resulting in:

> more informed and insightful decisions on actions to reduce their impacts, such that in 10 years, when comparable events occur, there will be a reduction in loss of life, fewer people adversely impacted, and wiser investments and choices made by governments, the private sector and civil society.[8]

The new programme proposed a forensic, evidence-based approach, 'Forensic Disaster Investigations', later known as FORIN – Forensic Investigations of Disasters.[9]

How would this scientific approach enable an improved understanding of risk? Would this be capable of prediction, identifying ways that 'in 10 years, when comparable events occur, there will be a reduction in loss of life'?

The FORIN programme referred back to Susan Stonich's work 'I Am Destroying the Land'. She had conducted what might be seen as a forensic investigation of hazards, vulnerability and risk – the three elements of the equation. A FORIN report reproduced her diagram of the *progression of vulnerability* in Honduras, showing factors which contributed to pressures on the population, resulting in an increasingly impoverished rural population that responded in ways including urban migration and further agricultural exploitation which endangered human well-being.[10] The detailed diagram is reproduced in Figure 4.

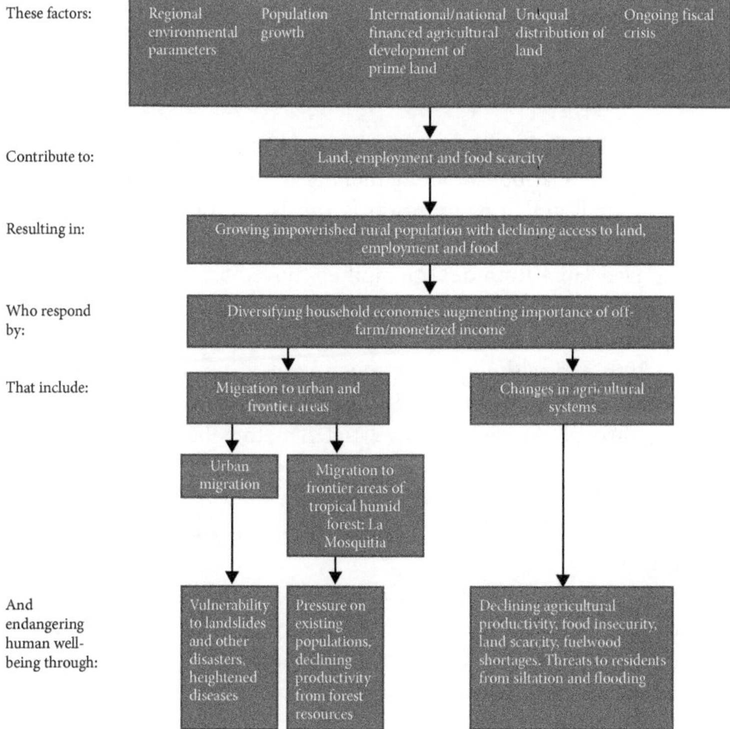

Figure 4 Interconnections between development, rural impoverishment and environmental destruction in Southern Honduras.
Source: Adapted from 'I am Destroying the Land!' Stonich. 1993.

The factors she identified and documented included large-scale agricultural development taking an unequal share of the land; population growth; an ongoing fiscal crisis leading to a struggling rural population with declining access to land employment and food; small farmers exploiting the little land they had access to; and migration either to increasingly overcrowded areas or to other previously untouched regions.

The bottom line of the diagram, and of her analysis, was that endangered human well-being resulted from *vulnerability* to ill-health, loss of livelihoods, food insecurity, landslides, downhill siltation and flooding. Her study was published in 1992, six years before the storm, and can be seen as 'forensically' predicting the disaster which struck the country when Hurricane Mitch swept through. Would her analysis reflect the reality of that awful tragedy?

Hazard, vulnerability and risk in Honduras: Mitch and the aftermath

Hurricane Mitch struck the country in October 1998: developing initially in the Caribbean, strengthening to hurricane force four days before landfall and moving towards the Honduran and Nicaraguan coast. It reached the coast of Honduras on the 27th of October, its eye resting over Santa Rosa de Aguan, then moved across the country. Its slow progress, dumping huge amounts of rain along its track, accounted in part for the extreme devastation it caused. It reached Tegucigalpa on the 30th and, although now much weaker, the rainfall combined with mud and landslides caused the Choluteca and Chiquito rivers running through the capital to swell to ten metres above their riverbeds. Torrential rainfall and floodwaters triggered a sudden landslide which swept a whole residential area, Colonia Soto, away in minutes. That and other landslips clogged and dammed the river, causing even greater flooding.[11]

The hard facts of the disaster are that eleven thousand people died, and many more were recorded as missing. About 76.6 per cent of the population were affected by the storm through loss of houses, land and livelihoods. The infrastructure and agriculture of the country were severely damaged, 70 per cent of the national highway system needed repair or rebuilding and ninety-two bridges were destroyed. More than 29 per cent of the country's arable land was affected; 85 per cent of banana, 60 per cent of sugar cane and 58 per cent of corn production was lost. A quarter of all schoolrooms and over 10 per cent of all health facilities were lost. Economic growth in the country slumped from an expected 5 per cent in 1998 to zero, and to -2 per cent in 1999.[12]

The inhabitants of Nueva Suyapa on the high edge of Tegucigalpa, their mud streets and paths washed away, suffered, like the rest of the city, from power outages and loss of water supply. They looked out on the shattered city, buildings flooded or even washed away, roads damaged and destroyed, bridges collapsed, the whole city at a standstill.[13] In Santa Rosa de Aguan villagers endured a battering of three days as the storm ripped roofs off their houses and the sea engulfed others, until an even greater disaster struck suddenly as the lagoon broke its banks and a flood cut right through the centre of the village, ripping it in half. Of the five hundred families in the village ninety lost their homes, and forty-two people died.[14] In the Olancho region deforested farming areas were flooded and, in some cases, washed away, destroying arable crops, while forested areas were heavily affected by treefalls. The most badly affected were the poorer farmers who had been pushed onto sloping, unstable land.[15]

Stonich's forensic analysis of the factors in Honduras magnifying hazards and vulnerability, including degradation of rural and urban environments resulting from poverty and exclusion, leading to endangerment from landslides, downhill siltation and flooding, proved sadly accurate. Storms are a characteristic hazard of the region, but the scale of the disaster which unfolded in October 1998 resulted from the growing *vulnerability* and *hazard* exposure of the country's population.

Attempting to be objective and keeping one's scientific hat on, one might question this analysis and ask whether the outcome would have really been much different if the population had been less vulnerable and less exposed to hazards. After all the storm was a powerfully destructive event. Was Stonich correct in her analysis of causes and effects? Did the differing vulnerability of sectors of the population significantly affect the outcome? In the 'laboratory' of human society it's more difficult to answer such a question than in a science lab, where control experiments can easily be conducted to compare outcomes. One test of Stonich's analysis might be to compare the outcome of the storm in a highly vulnerable region with that in a region not suffering the vulnerabilities she describes. We have seen how Susman and his colleagues did something similar, comparing Hurricane Fifi and Hurricane Tracy to see how their differing vulnerability affected outcomes.[16]

In the case of Hurricane Mitch we can look at an area of Honduras more protected and less exploited than the rest of the country – and therefore influenced less by magnified hazards and vulnerabilities – Mosquitia. We've already seen that this corner of the country still retained much of its native vegetation and that its population consisted largely of indigenous groups practising their traditional lifestyles, despite pressures of inward migration and logging. Would they, as a result, be less vulnerable to the impact of the hurricane?

One assessment comes from Kendra McSweeney, who conducted long-term research in Mosquitia, focusing particularly on Krausirpe, Isidro's village, from 1994 to 2002, spanning the period before and after the hurricane. In a paper focusing on the consequences of the disaster for the village and the surrounding area she reached conclusions which neither completely confirm nor confound Stonich's analysis.[17] On the one hand they showed that the storm caused considerable disruption and suffering despite the region's remoteness from the vulnerabilities Stonich identified. On the other they showed that the outcome of the disaster over the next four years was surprisingly positive.

When the storm hit Krausirpe over thirty centimetres of rain fell in three days. The Patuca river running past it rose by a similar amount, wiping out most of the rice, banana, plantain and manioc crops and leading to hunger and illness. Nevertheless, the undamaged state of the watershed and surrounding land meant this rise was far less than the *ten-metre* rise of the Choluteca river reported in Tegucigalpa, where deforestation and the built environment magnified the effects of the rainfall.

In Krausirpe, where there had been a push from local CSOs to concentrate on cacao production, 95 per cent of the cacao orchards were lost. Vulnerability was increased further by the traditional landownership system which favoured older established families and left others marginalized and unable to develop viable farms. Both these problems were amplified, ironically, by CSO and state efforts to conserve the forest, which had pushed the villagers further towards specialization on the now-destroyed cacao production. The picture McSweeney paints in the immediate aftermath of the storm is not encouraging. Krausirpe, just like other supposedly more vulnerable areas of the country, suffered substantial setbacks to the livelihoods of its population.

By the end of the study period, four years after the storm, the story was very different however. The village and its surrounding area had always been led in what McSweeney described as a 'bottom up, almost viral way'. She associated this and a generally high degree of social cohesion and cooperation with movements to correct the vulnerabilities that had accumulated before the storm. Villagers turned away from the specialization on cacao and re-diversified their agriculture. They moved onto new land, under the forest canopy, and rapidly changed landownership practices so that those previously marginalized were able to access parcels of land and build livelihoods. Though not driven by formal conservation measures, the new approach reduced forest clearance as farming was increasingly conducted under the forest canopy.

McSweeney monitored the response of the village to subsequent storms – Tropical depressions sixteen and forty-three in 2008 – and found that the changes in the village economy resulted in much lower impact from the storms. The villagers avoided farming the floodplain areas, and the safer areas they did farm were undamaged. They had accumulated capital which allowed them to cope with the short-term effects of the storms. They had secured sources of clean water through a community project to create a reservoir. Reflecting back on the 'Access Model' they experienced a virtuous, rather than a vicious circle in which they had improved their ability to cope with shocks and stresses. Response to Mitch in the area seemed to reflect the often-used phrase 'build back better'.[18]

The experience of Krausirpe suggests that while an intensive natural hazard on the scale of Hurricane Mitch will have substantial immediate impacts regardless of the level of vulnerability of a population, its ability to recover, 'bounce back' and build back better may be determined by its degree of vulnerability and hazard exposure. In the case of Krausirpe the *environment* was relatively intact, compared with much of the country where deforestation was substantial. The *social structure* was stable and appeared cohesive, rather than being fragmented and disrupted. These factors helped in re-establishing agriculture, avoiding the flood plains and reorganizing social and economic structures. They also showed a turn away from externally imposed specialization into cacao production.

If a less vulnerable population is able to 'bounce back' – building back better – as appeared to be the case in Krausirpe, would the same be the case in other regions of the country where vulnerability as a result of environmental degradation, economic inequality and social upheaval were greater? What of the villagers of coastal Santa Rosa de Aguan, of the ranchers of Olancho or the inhabitants of the Nueva Suyapa shanty community on the edge of Tegucicalpa? What of the country as a whole? While the hard facts of the disaster's impacts are stark, did these other localities also 'build back better' or – as Stonich suggested – did the vulnerability of these regions erode their capacity to weather the storm as Krausirpe had? Did the equation work?

Transformation?

Initial signs were hopeful. In the wake of the disaster donors poured funds into the country to support recovery and reconstruction. Funding from individual countries including France, Germany, Italy, Japan, Netherlands, Spain, Sweden, the United States and other smaller donors totalled just over $1 billion. A further $1.6 billion came from international (multilateral) institutions.[19] The scale of funding led to calls for coordination. The 'Consultative Group for the Reconstruction and Transformation of Central America' was established a month after the hurricane. The Swedish development agency, SIDA, a substantial donor, pressed for funding goals to reflect principles of *reconstruction* and *transformation*. 'Transformation', in 'development-speak', means doing more than just addressing short-term needs. It means 'building back better', reducing vulnerability to future shocks.[20] At the second meeting of the consultative group in Stockholm, in May 1999, the scope of this transformation was agreed and enshrined in the 'Stockholm Declaration'. Here are the goals set out in that declaration:

- Reduce the social and ecological vulnerability of the region, as the overriding goal.
- Reconstruct and transform Central America on the basis of an integrated approach of transparency and good governance.
- Consolidate democracy and good governance, reinforcing the process of decentralization of governmental functions and powers, with the active participation of civil society.
- Promote respect for human rights as a permanent objective. The promotion of equality between men and women, the rights of children, of ethnic groups and other minorities should be given special attention.
- Coordinate donor efforts, guided by priorities set by the recipient countries.
- Intensify efforts to reduce the external debt burden of the region.[21]

These were ambitious goals! They called for substantial transformation of Honduras compared with its condition before the storm. We have seen that the country's social and ecological vulnerability, the first goal to be addressed by the declaration, had been increasing. Transparency and good governance were woefully poor as were democratic standards, degrees of decentralization and involvement of civil society.[22] Human rights standards were poor, and the final goal of reducing external debt was far from being achieved, in fact much of the income intended towards reconstruction was being diverted to servicing existing international debt.[23]

It seems hard to imagine the country changing course and achieving these goals over even a relatively long-term time frame, let alone over the four-year period set by the declaration. Assessments conducted by a World Bank-funded group and also by the Swedish SIDA development agency at the end of this period concluded sadly, but predictably, that these goals had not been achieved.[24] Looking at the immediate reconstruction of Honduras, they found donors often set their own goals for use of their funding. There was little national coordination so money was not spent responding to the most important needs, but on what donors felt was most important. Much of it aimed to achieve short-term goals rather than longer term change. Pressure to spend funds fast so that donors could report results meant that work was rushed and often of a poor standard. In rural areas funding focused on restoring the status quo with its continuing trend towards deforestation, while in urban areas unemployment remained high, with no initiatives to create new jobs. The study concluded that 'Transformation had not taken place'.[25] The leader of a Honduran NGO network said that in his opinion it was all talk:

> There is discussion now on the causes of poverty, while there had not been any before. This is new. This consciousness on the part of ordinary citizens is new. But this is still all at the level of debate, and discussion.[26]

From the donor perspective one representative of a funding country said they might just throw up their hands and go:

> Reconstruction more-or-less happened, but transformation has not. Security has deteriorated dramatically, poverty is increasing. The [slump in prices leading to the] coffee crisis is more devastating than the drought. If we [donors] don't see fundamental transformation we shall leave.[27]

Looking to the future, a local expert said he detected a vicious circle:

> The disasters cycle is getting shorter, we are more vulnerable, and losses are higher. The location of human settlements is increasingly vulnerable.[28]

The conclusions of these reports were borne out in the communities we've previously encountered – Santa Rosa de Aguan, the Olancho farming region, and Nueva Suyapa and other shanty settlements on the fringes of Tegucigalpa. None of them saw transformation towards a more secure future: In Santa Rosa de Aguan residents were still struggling with options for reconstruction which were tearing the small community apart, four years after the storm. Their original locality had been cut in half and shrunken by the river cutting through it and the sea eroding the dunes. Their two options were of either moving to a housing settlement being built inland on the access road, away from their former home and from the sea, or of moving to an area along the coastline – similar in character to Santa Rosa – at La Planada, but with no road access inland. They felt that in either case they would lose their community and their culture.[29]

Inland in the Olancho region, destruction of crops, dwellings and damage to farmland had made farmers poorer. Increased migration, predominantly of the younger generation, created a labour shortage. A push towards milk production supported by the United States was increasing pressure for further deforestation, and others were turning towards carpentry to generate income, which was additionally increasing demand for timber. The downward trend in the quality of livelihoods and the state of the environment continued.[30]

The capital city Tegucigalpa and the poor informal communities fringing it displayed visibly the findings of studies and social surveys that peoples' lives had got worse, not better.

> While macro-indicators provide mixed conclusions, interviewees believe they are worse off now than prior to the disaster. Interviewees at all levels reflected this conclusion. ... In the urban sector, no initiatives for employment creation were found.[31]

Walking down a main street in the city while working there in 2002 I saw an ice cream truck with an armed guard hanging off the back of it. Social pressures, unemployment, poverty and gang warfare made even ice cream a valuable commodity. Crime trends, an indicator of poverty and social instability, were heading inexorably upwards.[32]

Marcos Burgos, commissioner of the national emergency response centre, pointed out, ten years after the storm, a bridge over the Choluteca river which runs through the city centre.

> You see that? That's a Bailey bridge. It was put up 10 years ago. It was supposed to be temporary but it's still being used. It's fallen down three times because people have stolen the screws.[33]

He also pointed out a gleaming new shopping centre, built in just seven months. The site was that of the Trebol refugee camp which had housed three thousand people. Those residents had to wait three years to be finally moved out of their tents. He says that much of the rebuilt infrastructure was based on out-of-date, risky criteria and would not survive another storm. Hector Espinal of UNICEF said:

> If the state suffered another Mitch, more people would definitely die. The country hasn't learnt its lessons. The state institutions still don't have the capability.[34]

Nueva Suyapa hit the headlines in the United States in 2003 when a *Los Angeles Times* reporter, Sonia Nazario, told the story of a boy called Enrique fleeing the settlement, desperate to escape drugs-driven violence in his community and even at school. The article expanded into a widely read series and book, telling Enrique's story and also accounting for the fact that by 2010 over twenty thousand children per year were fleeing places like Nueva Suyapa in Honduras, driven out by brutal, violent crime and gang wars.[35]

Transformation seemed, according to assessments and surveys, to have stalled, the promises of the national government and international community in the Stockholm Declaration long forgotten. Standing in a half-finished rehousing scheme, resident Nahum Cáceres fished out a piece of paper from his wallet on which he'd kept notes of the dozen international aid organizations that had come and gone since the disaster. He said:

> I don't know how much they sent, but they tell me this is a million-dollar project. I would like them to see what has happened with all their money.[36]

Never-ending disaster rather than transformation

In the wake of Mitch, as years passed, it slowly became apparent that the disaster hadn't gone away. Deforestation, corruption, exploitation, marginalization, failing infrastructure, precarious employment, social disintegration: none of these had been addressed or dispelled. The whole history of Honduras pre- and post-Mitch is one of accumulating risk in the decade (and many more years before) preceding the storm, and of a failure to 'transform' the economy and society of the country in the wake of it, simply recreating the conditions which existed before. What to the outside world was a sudden and overwhelming disaster event was in fact a spike in a history of 'never-ending disaster', a grinding, *everyday* pattern of stresses and shocks to lives and livelihoods.

I've mentioned my increasing awareness of 'everyday disasters' – AIDS infection in rural Tanzania, seasonal flooding in a Benin suburb, repeated droughts on the Eritrean border. Everyday disasters like these were a major focus of GNDR's VFL: large-scale social surveys of people in many lower- and middle-income countries, asking them about the disaster events, large and small, that affected them. While I was working on this at GNDR from 2008 to 2016, eighty-five thousand people in urban and rural communities were consulted.[37] We found that the threats that most concerned people included flooding, drought, insecurity, earthquakes and landslides, disease and epidemics, and pollution. The *consequences* of these threats were in many cases losses and damage rather than injury, death and displacement. They reflected small-scale everyday disasters[38] (see Figure 5).

We also found that when people, self-assessing their economic status, were asked whether conditions were worsening or improving, the poor felt that conditions were worsening, whereas those who were much better off thought conditions were improving; in other words, 'the poorer you are the worse it gets' (Figure 6).[39]

The emphasis in 'Views from the Frontline' on everyday disasters was increasingly acknowledged. The UN's own research found that losses from what they referred to as 'extensive disasters' were outpacing those from headline 'intensive' disaster events.[40] Whatever the term, the message was the same, that for the majority of the populations of Honduras and other LMICs disasters were not occasional intense events, they lived in a condition of never-ending disaster.

'Views from the Frontline' surveys showed a connection between poverty and vulnerability to everyday disasters. Vulnerability and

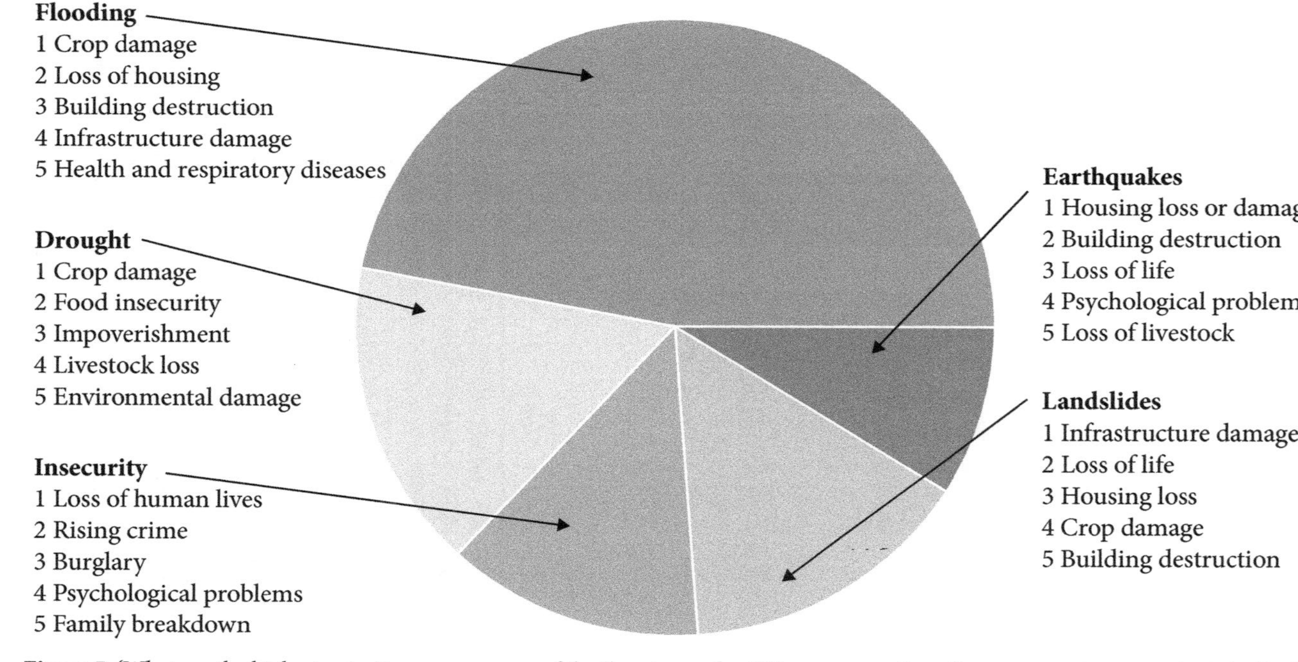

Figure 5 'What are the highest priority consequences of the threats you face?' Responses to Frontline community survey in Asia, Pacific, Africa and Latin America/Caribbean (n = 14,282).
Adapted from 'Frontline: from local information to local resilience'. GNDR. 2015. Oliver-Smith et al. 2017.

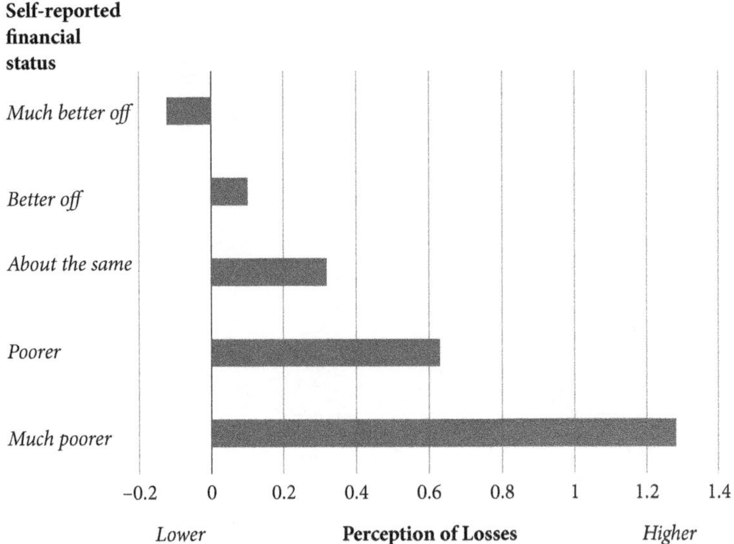

Figure 6 Responses to survey question 'Are losses increasing or decreasing?'.
Source: Adapted from GNDR 2013.
Note: Based on 21,455 responses from fifty-seven countries. 'Perception
of disaster' losses axis based on Likert scale. 2 = Substantial increase,
1 = Slight increase, 0 = No change, −1 = Slight decrease, −2 = Substantial
decrease.*Source*: Oliver-Smith et al. 2017.

everyday disasters are closely related. Factors which make people
vulnerable – limited choices and limited resources resulting from
inequality – expose them to a grinding succession of unavoidable
everyday disasters as their houses flood in the seasonal rains, disease
spreads through cramped housing with poor water and sewage
infrastructure, fires easily start and spread with poor services to prevent
or respond to them, and so on. Everyday disasters, as a symptom of
vulnerability, sadly reveal who is likely to be most affected when a major
disaster strikes. In Chapter 1 I mentioned Amar, the Sudanese farmer
sitting with the last of her five hundred goats after the rest had been
gradually sold off as every drought hit. Having lost all her resources,
Amar was frighteningly vulnerable to the next disaster striking their
village. This is the harsh reality of the progression of vulnerability and
of everyday disasters.

Vulnerability in Honduras: Root causes – corruption

Why, despite the scale of the disaster and the massive commitment of resources in the weeks and years following it, did the country and the capital build back, not better, but the same as before … if not worse? The answer coming from both within the country and from outside commentators is clear. Corruption driven by powerful interests within government and big business, much of it multinational, maintained the exploitation and destruction of the natural environment and the impoverishment of the population, increasing the risk of further disasters. A study published in 2020 found four kinds of corruption siphoning the country's resources away from effective development and social progress, into the pockets of officials and business:[41]

- Embezzlement of social development funds: Transferring public funds to 'shell' non-government organizations, supposedly to carry out social development projects, but in fact to hollow entities created to siphon funds into personal, campaign and party accounts.
- Procurement corruption: Government officials used false procurements to embezzle funds and directed contracts to preferred suppliers at inflated prices in exchange for bribes and political contributions.
- Organized crime corruption: The police have been the primary target of organized crime corruption and several former members of Congress have been convicted in US courts for involvement in organized crime. Testimony in US courts produced allegations that the last three elected presidents took major campaign contributions from criminal groups.
- Patronage abuses in the bureaucracy: Incumbents have worked to increase their discretion over hiring and promotions as a source of patronage for campaign workers. In addition, incumbents collect donations from many public sector employees through automatic pay check deductions.

The report cited the example of emergency procurement of mobile hospitals during the Covid-19 pandemic at a cost of $47 million from a US-based shell company at a multimillion-dollar markup. The company was paid 100 per cent upfront with no guarantees or penalties. The head of the purchasing agency who struck this deal was eventually charged and sentenced to eleven years in jail.[42] Notice that in this, as in many other cases, corruption was not just within the country but on the part of businesses and organizations from the United States and other countries.

The organization which unearthed this violation was ASJ ('La Asociación para una Sociedad más Justa' – Association for a More Just Society) which was founded in Honduras in 1998 with the aim of tackling corruption and injustice in the country. They assembled a team of lawyers to pursue their cases and tackle exploitation by local businesses and international companies, help victims secure justice for unsolved crimes and investigate corruption cases within government – such as the hospital case.[43]

Why is it that ASJ employed lawyers, rather than climatologists or aid workers, to make Honduras a safer place? How can the underlying drivers of disasters be tackled by addressing injustice and corruption? This seems a very different approach from that of predicting and responding to disasters based on establishing early warning systems, orchestrating rapid response and managing large-scale reconstruction. The implication is that vulnerability is entangled with social and political processes, that risk has a social dimension, that it is created and magnified within our social structures rather than externally generated and imposed. If this is so, then how is risk created, and by who?

How is risk created? The social creation of disasters

Addressing exploitation and corruption degrading the physical environment, eroding livelihoods and disrupting social stability and cohesion were all targets of the failed attempts at transformation in the wake of Hurricane Mitch. None are the results of natural hazards. From the early recognition of the relationship between hazard and vulnerability it has been increasingly recognized that this relationship drives the *creation* of risk. Examples like that of Honduras shifted the focus further to recognize that risk is *socially constructed*.[44] While a major disaster event such as Mitch created massive environmental impacts, the outcome of the disaster was shaped by social factors. The established and cohesive communities of Mosquitia were able to recover and adapt, building back better. Rural communities on the coast and inland struggled as a result of the accumulation of risk. That risk was generated by social and economic pressures driving displacement and environmental degradation – the exploitation of much of the land by big agro-business, much of it multinational, and the squeezing of the population onto marginal land, needing to strip yet more land of forest to eke out a living. In the city the interests of big business and corrupt government diverted resources away from development, leaving much of the population struggling with poor housing, education and health services, fragile infrastructure, exploitative employment and increasing poverty driving crime and gang warfare. The risks the rural and urban populations experience are *socially constructed* by the actions of business and government.

The concept of *La construcción social del riesgo* – the social construction of risk – emerged in Latin America[45] representing a further step change in thinking about the roots of risk and the challenges of achieving change for the better. 'The social construction of risk' suggests that the interplay of different elements of society is responsible for the accumulation of risk within societies, and that vulnerability is not distributed evenly through them.[46]

The idea of 'never-ending disaster' afflicting the population of Honduras and the roots of this experience in political, economic and social pressures reflect this social construction of risk, emphasizing an idea which has been evolving throughout this chapter and throughout the history of disaster studies. It highlights the role of human actions or 'agency' in creating risk. While this may seem hard at first to understand when confronted with the intense impact of Hurricane Mitch over just three days, broadening the focus to the pre- and post-disaster story makes it easier to see the experience of small-scale, everyday 'never-ending disasters' as the consequence of human choices. This is underlined by experiences analysed in VFL and by work done by agencies such as ASJ. Everyday disasters are the consequence of everyday risk, and this risk is created as a result of choices made by those with the freedom and power to make them. In Honduras the best land, whether for farming or housing, is secured by powerful and rich interests and as a result poor farmers are squeezed onto marginal land or, worse, are forced to 'destroy the land' as the Honduran farmer said, knowing that their actions are degrading the environment but having no choice. In cities prime sites are secured by big businesses and rich residents, pushing poorer residents onto cheaper plots – cheap because they are unstable or low-lying, vulnerable to floods, erosion and landslips. Large businesses squeeze labour rates for poor workers, while corruption in government siphons funds meant for social protection and infrastructure into the pockets of the powerful. The environment is degraded and destroyed by exploitation for large-scale farming, mining, logging and on an international scale by fossil fuel extraction and carbon emissions contributing to global warming.

What Stonich showed in her analysis of the conditions developing in Honduras before Mitch was that the creation of *everyday risk* accumulated with the result of intensifying the disaster impact of the storm. In other words, creation of risk proceeds through an accumulation of everyday risk which not only results in persistent everyday disasters but also creates conditions for a major trigger event to result in massive disaster impacts.

Chapter 4

HAZARDS

What about hazards?

The equation lying behind much of our discussion places a strong emphasis on vulnerability. The causal chain which the 'crunch' diagram depicts shows this sequence increasing exposure of vulnerable groups to hazards. But what of those hazards? Are hazards themselves fashioned and amplified by human action? Are they not only 'natural' but also 'technological' and 'socio-natural' as the 'FORIN' project suggests?[1] Are hazards socially created, just as vulnerability is? Can the stories of the lives of people in Santa Rosa de Aguan, the Olancho region, in urban shanties such as Nueva Suyapa and even increasingly in Mosquitia be seen as impacted by never-ending disasters resulting in part from increasing hazards, not just 'natural' but amplified by social and technological choices?

Chernobyl and beyond: 'Risk Society'

On 26 April 1986 a safety test in reactor number 4 of the Chernobyl nuclear power plant in north Ukraine led to loss of control of the reactor core, causing a reactor meltdown and explosion. The building was destroyed and radioactive material was scattered over thousands of kilometres across the USSR and Europe. The stricken reactor was later sealed under a protective sarcophagus and stands at the centre of a 'zone of alienation' – an exclusion zone of 2,600 square kilometres – as a memorial to the event. In the year of the accident Ulrich Beck published 'Risikogesellschaft' which – translated and published in English six years later as 'Risk Society' – painted a picture of society under the shadow of manufactured and magnified hazards.[2] For Beck the growth of technology over the previous two centuries had created a new risk landscape in which society had to find 'a systematic way of dealing with hazards and insecurities induced and introduced by modernisation itself'. Chernobyl was a powerful example of a hazard born out of modernity.

Quite apart from disaster events such as the Chernobyl meltdown, society had been living under another, darker shadow, the threat of a nuclear apocalypse. The era was darkly celebrated in Stanley Kubrick's 1964 black comedy, 'Dr. Strangelove', in which the 'Doomsday Machine' guaranteed 'Mutually Assured Destruction'. Black comedy it may have been, but the 1962 Cuban Missile Crisis which preceded the film threatened a real-life apocalypse as the USSR sited nuclear weapons within striking distance of the United States, in retaliation for their siting of nuclear weapons in Italy and Turkey.

Beck's focus was on hazards, their creation and amplification in modern society. This is a very different emphasis from what we've seen so far. Hazards are no longer a 'given' as they are in the 'crunch' diagram, in which all the action is on the vulnerability side. In his thinking, the industry which creates our modern world also manufactures and magnifies hazards.

Are hazards being magnified?

'At Risk' – the 1994 work which presented and unpacked the crunch diagram – took note of Beck's thinking.[3] However, with an emphasis on the lives of risk-exposed people in less developed countries it felt that Beck's focus on hazards driven by technology and modernity was less relevant than peoples' vulnerability to a range of well-known hazards such as flooding, storms and earthquakes. Leaving aside the shadow of nuclear Armageddon, which was global, the lives of people in places such as rural Tanzania, for example, were far more impacted by storms, droughts, famines and disease than by Beck's technological hazards. But beyond the Cold War there were other ways the rich, high-tech Global North could ripple hazards out into the South, as I saw in 2008.

I spent some time in Tanzania and neighbouring Zambia that year, gathering information about the impact of debt and debt remission in those countries. As I did so another shadow was sweeping across the world. The 2007–8 financial crash was a disaster on a massive scale, wiping trillions of dollars off global balance sheets.[4] It was the consequence of trading in high-risk investment vehicles and high-risk loans. It rocked the developed world, but the shadows spread way beyond Wall Street. People I talked to in cities, towns and villages along our route through those two countries found not only that fuel prices had shot up but that even the food they bought daily had more than doubled in price. The whole world had got poorer. Tanzania and Zambia had become part of Beck's 'Risk Society'.

While *vulnerability* to hazards results from social and economic inequality, placing millions in conditions of everyday disasters and increased vulnerability to intense disasters, perhaps the *hazards* they face are being magnified or even created through human activity. It is unarguable that new hazards such as those resulting from nuclear energy and weaponry have been created, and the financial crash of 2007–8 suggests there are other ways in which hazards are being magnified; but is society more generally creating and magnifying the hazards people are exposed to? Consider two examples of disaster events and the hazards which lie behind them:

Climate change: Hurricanes and typhoons and their consequent structural damage, flooding, injury and loss of assets are *natural* hazards, but these climate-driven hazards are *magnified* by increasing carbon emissions leading to global warming, predicted to lead to more frequent and intense weather events. Magnification of climate hazards is therefore a consequence of growth of industry, manufacturing and travel.[5]

The Turkey/Syria earthquake of February 2023 was the result of a hazard. Earthquakes are seen as uncontrollable, unpredictable events. In this case Turkey's President Erdogan described the disaster as a case of 'force majeure'[6] – beyond any possibility of the government anticipating it. But as I mentioned in Chapter 1 the hazard had been *magnified* by failure to adhere to planning laws designed specifically, after the previous earthquake, to anticipate and reduce the earthquake hazard.

Beck's focus was on new hazards resulting from technological change and progress. However, much human activity has the effect of magnifying *existing* hazards. Climate scientists regard global warming as magnifying the likelihood and intensity of extreme weather events, rather than creating new hazards (as nuclear power and weaponry do). There's a shift of focus here as it seems that human activity can magnify hazards in various ways, as in the case of the Turkey earthquake where the 'initial hazard' (a pre-existing hazard), in this case the earthquake, is *magnified* by poor and even illegal building practices leading to the *realized hazard* – unnecessary building failures.

Looking at disasters resulting from different hazards and at different scales it seems there are many cases where *initial* hazards are magnified to create *realized* hazards. Taking the UN disaster agency UNDRR's classification of hazard types, Table 1 considers examples of each, suggesting the initial and realized hazard in each case:[7]

Growing hazards: The dynamic progression of hazards

Whereas Beck's emphasis was on newly created hazards emerging from technological advances, such as the Chernobyl disaster, all the examples in Table 1 are of initial hazards which have been magnified by human actions, including the proximity of infected animals to human food, the intensification of vehicle emissions in growing cities, growing coastal populations on Japan's East Coast (the Sanriku coast which was most heavily affected was well known locally to be particularly susceptible to tsunamis[8]), unmanaged deforestation, irrigation and building in the Indus River basin, and failure to manage storage of explosive substances safely at Beirut Port. These examples suggest a dynamic progression of hazard, from *initial* hazard through hazard *magnification* to *realized* hazard, and suggest that rather than hazards being simply natural phenomena they are magnified by human actions and choices, mirroring the concept of a progression of vulnerability in the 'crunch' diagram. Figure 7 (p. 55) develops the 'crunch' model to include this idea.

Since the publication of 'Risk Society' the idea of the creation and magnification of hazards through human action – anthropogenic hazard creation – has gained traction. This idea has moved in tandem with the awareness that this activity is changing the face of our planet. Scarred and deforested landscapes, smog-filled skies over cities, polluted rivers, acidified and warming oceans, changing weather patterns, species loss – all these have led to suggestions that we are entering a new geological age – the 'Anthropocene' – in which human activity makes a significant impact on the earth's geology and ecosystem.[9] It flags recognition that the world is increasingly affected by human activity, in doing so both magnifying and creating hazards. Whereas in a pre-industrial era hazards were largely environmental (with the exception of violence and war) the Anthropocene reflects the emergence of an array of dynamic causes of creation and magnification of hazards. In considering unnatural disasters it is increasingly clear that we need to recognize the social creation of *hazards*, alongside the social creation of *vulnerability*.

Table 1 Hazard Types and Examples of Initial and Realized Hazards

Hazard type (UNDRR)	Example	Initial hazard	Hazard magnification or creation	Realized hazard
Biological hazards include bacteria, viruses, parasites, poisonous plants and animals and the carriers of infections, through air, water and animals such as mosquitos.	**Covid-19**	Disease in animals, limited human hazard	Zoonotic (animal-human) transfer likely. Resulting from lack of regulation of wet markets, wildlife trade and farming practices[a]	2020 Covid-19 pandemic
Environmental hazards include chemical, natural and biological hazards resulting from environmental degradation (soil degradation, deforestation, loss of biodiversity, salinization and sea-level rise), such as physical or chemical pollution in the air, water and soil.	**Urban air pollution**	Limited pollution in pre-industrial cities	Increasing motorized transport producing particulates, CO, NO_2, SO_2 and VOCs. Concentration of pollutants in growing urban concentrations	One study finds 3.5 million premature deaths worldwide in 2017 attributed to transportation air pollution[b]
Geological or geophysical hazards originate from internal earth processes. Examples are earthquakes, tsunamis, volcanic activity and emissions, and related geophysical processes such as mass movements, landslides, rockslides, surface collapses and debris or mud flows.	**Great East Japan Earthquake and Tsunami 2011**	Coastal flooding and destruction of built environment	Urban development in exposed coastal areas with inadequate sea defences. Location of Fukushima nuclear power plant with inadequate flood defences[c]	Magnified destruction of property and loss of lives and livelihoods. Nuclear accident and pollution
Hydrometeorological hazards are of atmospheric, hydrological or oceanographic origin. Examples are tropical cyclones (also known as typhoons and hurricanes); floods, including flash floods; drought; heatwaves and cold spells; and coastal storm surges.	**Indus River basin floods, Pakistan, 2010**	Monsoon rains leading to flooding of lowland areas in Sindh valley	Deforestation of upstream areas, siltation, unmanaged development of the built environment in lowland areas, failures of watershed management[d]	Floods affected more than 18 million people, caused 1,985 deaths, and damaged or destroyed 1.7 million houses. They were the worst floods in Pakistan's modern history.[e]

(continued)

Table 1 (continued)

Hazard type (UNDRR)	Example	Initial hazard	Hazard magnification or creation	Realized hazard
Technological hazards originate from technological or industrial conditions, dangerous procedures, infrastructure failures or specific human activities. Examples include industrial pollution, nuclear radiation, toxic wastes, dam failures, transport accidents, factory explosions, fires and chemical spills.	**Beirut ammonium nitrate blast, 4 August 2020**	Ammonium nitrate (AN), an ingredient of industrial and military explosives, is hazardous if not properly stored.	2.7 kilotons of AN was stored for six years at the port despite repeated safety warnings. Fireworks were stored adjacently and storage was next to the country's main grain store.	A fire ignited fireworks and this led to the AN explosion, described as the biggest non-nuclear urban explosion. It resulted in over two hundred deaths, three hundred thousand were made homeless and much of the country's grain reserves were lost.[f]

Note: [a] Lawrence O. Gostin and Gigi K. Gronvall, 'The Origins of Covid-19 – Why It Matters (and Why It Doesn't)', *New England Journal of Medicine* 388, no. 25 (22 June 2023): 2305–8, https://doi.org/10.1056/NEJMp2305081.
[b] Awais Piracha and Muhammad Tariq Chaudhary, 'Urban Air Pollution, Urban Heat Island and Human Health: A Review of the Literature', *Sustainability* 14, no. 15 (January 2022): 9234, https://doi.org/10.3390/su14159234.
[c] Ishigaki et al., 'The Great East-Japan Earthquake and Devastating Tsunami', April 2013.
Source: Author. (Based on classification from UNDRR. 2021.)
[d] M. Oxley, 'Pakistan Floods: Preventing Future Catastrophic Flood Disasters', *Preventionweb*, 2010, https://www.preventionweb.net/files/15697_01.10.101.pdf.
[e] 'Heavy Rains and Dry Lands Don't Mix: Reflections on the 2010 Pakistan Flood', Text.Article (NASA Earth Observatory, 6 April 2011), https://earthobservatory.nasa.gov/features/PakistanFloods.
[f] Samar Al-Hajj, Hassan R. Dhaini, Stefania Mondello, Haytham Kaafarani, Firas Kobeissy and Ralph G. DePalma, 'Beirut Ammonium Nitrate Blast: Analysis, Review, and Recommendations', *Frontiers in Public Health* 9 (2021), https://www.frontiersin.org/articles/10.3389/fpubh.2021.657996.

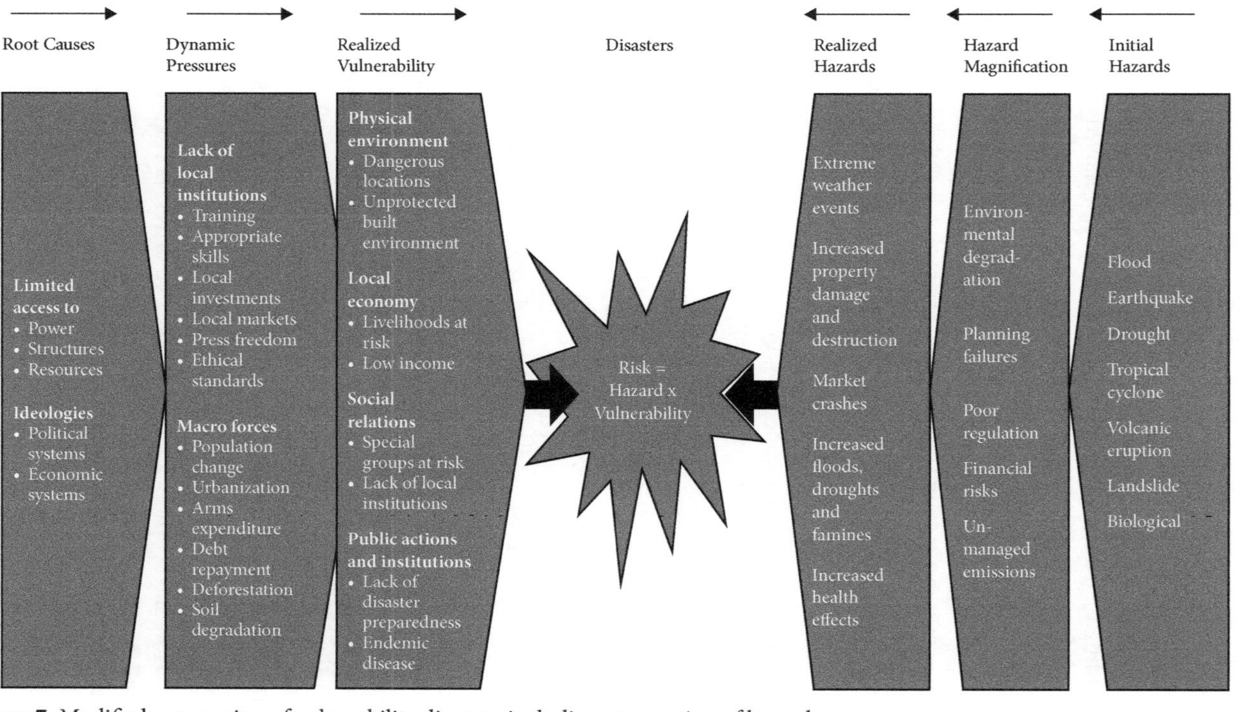

Figure 7 Modified progression of vulnerability diagram including progression of hazard.
Source: Adapted and simplified from *At Risk*, Blaikie et al. 1994 and *The Routledge Handbook of Hazards and Disaster Risk Reduction* Ch. 3. Wisner et al. 2012.

Socially created vulnerability, hazards and risk

Our consideration of hazards suggests that just as there is a 'progression of vulnerability' so there is a 'progression of hazards' and that human activity magnifies initial hazards (and in some cases creates them) leading to greater realized hazards. We have now moved even further away from the idea of disasters as inescapable and uncontrollable events, discovering that risk lies at the crunch point between vulnerability and hazards. As we've followed the chain of events preceding the Hurricane Mitch disaster in Honduras and seen what happened in the wake of the storm, we've found that a 'progression of vulnerability' affects the impact of a disaster such as this, and also influences the outcome and the possibility of 'transformation' – building back better. We're now suggesting that just as vulnerability is socially created, so too is much 'realized' hazard. While hazards of various types have always been with us, human activity magnifies and, in some cases, creates hazards. Both vulnerability and hazards are socially created and therefore risk is a socially created phenomenon.

The UN disaster reduction agency, UNDRR, conducts deep research on disasters and risk to underpin its coordination of international disaster risk reduction frameworks. In its 2015 flagship 'Global Assessment Report' (GAR) the UNDRR made the headline claim that 'New risks have been generated and accumulated faster than existing risks have been reduced'.[10] It continued to reiterate this in subsequent reports, right up to 2022.[11]

The year 2015 was the one in which the UNDRR launched its new fifteen-year disaster risk reduction framework, the 'Sendai Framework for Disaster Risk Reduction' (SFDRR), but its message seemed to be that the disasters trend, and the underlying risk driving it, was going in the wrong direction. The world, it seemed, was becoming a more dangerous, rather than a safer place. Risk creation was shadowing our lives and our future. For all the efforts of that organization and its member nations the pool of risk was increasing. Risk was being created, or magnified – or both.

How and by whom are *risk*, and the two drivers of risk, vulnerability and hazards, being created and magnified? Is there an underlying upward trend in vulnerability, hazards or both?

The report was the latest in a two-yearly series of substantial and widely researched studies. Its 2015 findings had additional weight, underpinning the finalization and launch of the SFDRR, replacing the 'Hyogo Framework for Action' which had been in place for the previous decade. This new report homed in on risk creation and on the consequences of current trends for the world's future. Adopting economic language it spoke about the 'mispricing' of risk:

> Currently, the absence of accountability in most societies in the face of both neglectful and deliberate risk generation means that consequences are rarely attributed to the decisions that generated the risks. At the same time, this lack of attribution creates perverse incentives for continued risk-generating behaviour. In effect, those who gain from risk rarely bear the costs. These costs are borne involuntarily by other social sectors and territories or transferred to the commons.[12]

The clear implication was that there were entities that were creating risk and offloading the consequences onto society. It went further to identify some of those entities:

> The inadequate pricing of disaster risk and of broader externalities in economic activity means that disaster risk is discounted excessively in order to maximize short-term gains. Moreover, there is no accountability for investments that generate disaster risks and are made by managers of large funds, banks, businesses and insurers, increasingly in cooperation with local and national governments.[13]

The report went on to look into the future consequences of current behaviour, introducing the idea of a new era under the title 'Welcome to the Anthropocene'. The chapter related risk creation to growing global overconsumption, leading, according to its analysis, to the world crashing through sustainable planetary boundaries:

> The ecological footprint currently created by this overconsumption of energy and natural capital now exceeds the capacity of the planet to provide the resources used and to absorb waste, including greenhouse gas emissions. Somewhere around 1970, consumption surpassed the planet's biocapacity for the first time. It is estimated that consumption now exceeds the biocapacity of the planet by around 50 per cent.[14]

What, exactly, is it that the report is talking about here? Vulnerability? Hazards? Or both? I suggest that what is being addressed here is *hazard creation and magnification*. The report refers to 'economic activity', 'investments', 'overconsumption', 'resources', 'energy'. It even (unusual for the UN) names names: 'managers of large funds, banks, businesses and insurers, increasingly in cooperation with local and national governments'. Their language, particularly the phrase 'In effect, those who gain from risk rarely bear the costs' reminds me of a comment from economist Joseph Stiglitz speaking of those responsible for the Great Recession of 2007–8. 'There's a system where we socialize losses and privatize gains.'[15] Unnatural disasters result from human choices: 'disaster risk is discounted excessively in order to maximize short-term gains'. This is an idea we will carry through our investigation of risk creation.

Focusing on hazards

I share the view of the UNDRR's Global Assessment Review 2015 writers and researchers that risk is being created at a frightening rate through the creation and magnification of hazards, the activities we've discussed above. Moving forward, it will be seen that these trends have become particularly dominant over the last seventy+ year period, termed by some 'The Great Acceleration'.

Governments however may still wish to maintain the fiction of disasters as unavoidable and uncontrollable rather than events for which they bear some responsibility – illustrated for example by Turkish President Erdogan's reference to the 2023 earthquake as 'force majeure'. If in fact disasters are not unavoidable and uncontrollable, but are *in* development and are a *consequence* of it, it gets more messy for governments who, as we will discuss later, are embedded in a system dependent on continued development and growth, whether or not this is sustainable.

Vulnerability may be increased or reduced depending on government stances on redistribution of wealth, provision of public services and social protection. These stances are largely politically determined. Right-wing governments emphasize personal freedom, shrinking the state, reducing taxes, reducing public services, accepting social inequality. Left-wing governments tend to increase redistribution, expand government, expand public spending and thereby increase public services and social protection. The latter are clearly likely to reduce vulnerability by increasing access. Levels of vulnerability at a national level are largely politically determined.

Hazard creation and magnification depend largely to the extent that the economic/industrial/consumption complex is managed. We will see as we move forward and consider this in detail that the lack of management has led to 'externalities' and 'market failures' which allow environmental degradation, overdevelopment, unmanaged urbanization, uncontrolled pollution and emissions. The mainstream economic response to these is that they should be dealt with by regulation. Regulation is therefore a recognized component, not only of managed governance systems but of 'free market'-oriented systems, in other words it ought to be exercised right across the political spectrum. The public expects (to the extent that it is aware of the problems and consequences) that it should be protected from negative consequences of the activity of the complex.

The drivers of vulnerability and of hazards are distinct and different. Increasing recognition of the magnification of hazards through burgeoning development and overdevelopment leads me to place a particular focus on investigating drivers of hazard creation and magnification, though in reality we will encounter, too, drivers of vulnerability along our journey.

Chapter 5

MANAGING DISASTERS AND DEVELOPMENT: THE UN FRAMEWORKS

Three frameworks, one goal

If, as we have seen, risk is socially created, rather than unavoidably part of our world, it's possible to take action to tackle the diverse causes of risk creation. Managing risk is the task of governments, and not only do governments grapple with their roles and responsibilities in dealing with disasters, but they participate in international frameworks such as the UNDRR's 'Sendai Framework for Disaster Risk Reduction' to coordinate action globally. While some disasters result from risk and vulnerability contained within national borders, many have an international dimension, resulting from events and actions beyond those boundaries. Referring back to Table 1, the tsunami which struck Japan's East Coast and the response to that were largely contained within Japan, but the Covid-19 pandemic did not respect borders. An emerging problem in Wuhan became everybody's concern. Even the case of major floods in Pakistan has an international dimension, as changing weather patterns are the result of global warming resulting from greenhouse emissions for which Pakistan has very little responsibility. Response also has an international dimension as both intense and everyday disasters often exceed national capacities. States often call for aid in times of major disasters, provided by the international system, CSOs and private donors. Everyday disasters often result from vulnerability as a consequence of poverty and poor infrastructure. Massive development aid flows attempt to address this. Recognition of the international dimension of disasters led to emergence in the 1990s of the first UN framework for disaster risk reduction. While the focus of this was initially narrow, concerned with intense disaster events, two other complementary UN frameworks, dealing with other aspects of risk, hazards and vulnerability, emerged over this period. The *climate* framework emerged from awareness of the potential effects of greenhouse gases on climate change, magnifying *climate hazards*. *Sustainable development* frameworks emerged from a concern to address *vulnerability* resulting from poverty. By 2015 these streams converged as three new key frameworks were established: The 'Sendai Framework for Disaster Risk Reduction' (SFDRR), the 'Sustainable Development Goals' (SDGs) and the 'Paris Agreement on Climate Change'. Though established separately it is widely acknowledged that these are three strands of a single cord, as the goal of *sustainable development* depends on driving down vulnerability and hazards through *risk reduction*, as well as addressing the particular hazard magnifier of *climate change*. All three frameworks underwent mid-term reviews in 2023 which we will examine later, as their language may promise much but – as with the Stockholm declaration to transform Central America in the wake of Hurricane Mitch – there may be a huge gulf between aspirational language and reality.

Lost in translation? The Sendai framework
for disaster risk reduction

These frameworks can be seen as attempts at 'global governance' of events which have international causes and consequences. They have their origins in the emergence of the UN system after the Second World War. While visible in times of humanitarian crisis and conflict as well as in developmental programmes, management of global environmental, social, finance and trade issues, the ability of the UN to exercise governance of any of these areas is limited. It may be more accurate to talk about *international* governance – an arena where nation states attempt to negotiate compromise or consensus which fits with their own priorities – rather than *global* governance in which some kind of global government is able to legislate on global issues. We will revisit this global governance issue more broadly in Part 3. However, the UN system, whatever its limitations, is what we've got, so in considering the management of risk, hazards and vulnerabilities it's important to understand what it can and can't do. Focusing on the SFDRR is a good case study of the way these frameworks are formulated and of what constraints limit them. What is true of SFDRR negotiations is similar in all UN framework negotiations, as they depend on securing the eventual agreement of all member nations with their widely differing priorities … as far as possible. I attended the 2015 world conference at Sendai, where the draft SFDRR, rooted in work such as the GAR studies, was to be finalized.

Japan, the host country, seems to have maintained a distinct and unique culture. Beneath global brand neon signs the invasion of global culture seemed far more limited than in other world capitals and reminded me of the film *Lost in Translation* – whose key protagonists were disorientated visitors to Tokyo encountering each other and experiencing a fragmented and fragile relationship within the cultural dislocation of the city.[1] The conference itself had a similar sense of cultural dislocation, mirroring the location, as people from many different nations playing many different roles converged on the centre. Politicians and statespeople from all member countries, CSO members and activists, scientists and researchers, bureaucrats and administrators all attended events, presented at them, peddled their wares in the marketplace and renewed old acquaintances in the halls, corridors and canteens.

Loss of vision?

Lost in Translation turned out to have another meaning at the event. The UNDRR's attempts to push understanding and action on disasters and risk forward seemed also lost in translation. Somewhere between the sharply critical analysis of UNDRR's 2015 GAR report and the framework which emerged, something got lost. This wasn't a total surprise as I'd learnt that the whole raison d'être of the network I was part of, GNDR, stemmed from frustration about a similar failure to translate analysis of the drivers of disasters into action, when the previous 'Hyogo Framework for Action' was agreed in 2005. Zenaida Willison, who helped establish GNDR, spoke out at that plenary on behalf of the civil society delegation to say:

> The translation of the ideals of the conference into political action is the task of the governments, but the responsibility of us all. … The outcome document does not reflect the spirit of the conference and the world around us. As it is now, it is a framework of vision and not a framework for action![2]

Just as Zenaida had seen in 2005, I found the 2015 conference had somehow lost the original thinking of the GAR in translation into its final outcome document. Around the Sendai conference halls and corridors thousands of copies of that GAR report were stacked ready for delegates. The report highlighted the idea that *new* risk was being created. This had not been part of the language of the previous frameworks.[3] But the process for finalizing and ratifying the new framework seemed to gradually disconnect from the sharp critical thinking of the GAR report. The negotiation process seemed shaped to guarantee this.

The heart of the negotiations

The conference itself had a strict hierarchy. Organizations like ours and many others had been able to book events within the programme ranging from large-scale presentations to smaller discussions and pop-up events, but the main events of the conference were under control of the UN secretariat and were restricted almost entirely to high-level delegates from national deputations, business and academia. These events were carefully choreographed, with limited and carefully controlled discussion slots. But there was another layer of hierarchy above this. The real heart of the conference was the ongoing negotiations to finalize the new framework – the outcome document of the conference. Zenaida describes this process at the preceding conference very well:

> Passionate statements were made in this plenary. New insights exchanged in the thematic sessions. But they have not reached the outcome document. In the dungeons of this conference, diplomats work until deep in the night on the outcome document. We see that governments are avoiding and eroding their responsibilities, instead of seizing this moment to make the strongest possible commitment.[4]

I'd read this description years before attending the 2015 event and now it seemed strangely familiar as we sat at various times, morning, evening and night in the small viewing gallery for 'observers' above a huge conference room, dotted with desks for each of the 187 national delegations. They were all wired up with microphones and simultaneous translation, viewing a massive screen on which the draft framework appeared, page by page. A convening group at the front negotiated changes and edits painstakingly, running overnight and way beyond the original deadline. As much as I could understand it the general approach was one of each delegation arguing for removal of statements they found problematic in terms of their own national priorities and pressures. The consequence was that, as Zenaida had described, it seemed to be a process of *avoiding and eroding their responsibilities*. The finalized framework set seven targets for the framework period from 2015 to 2030, summarized here:

To substantially *reduce* (compared with the previous framework period from 2005 to 2014):

- Global disaster mortality
- Number of affected people globally
- Direct disaster economic loss
- Disaster damage to critical infrastructure and disruption of basic services

And to substantially *increase*:

- The number of countries with national and local disaster risk reduction strategies
- International cooperation to developing countries (financial support)
- Availability of multi-hazard early warning systems and disaster risk information

Targets for substantial reduction in mortality, affected people, losses and damage are underpinned by targets to promote disaster risk *reduction*, finance to developing countries and early warning systems. But what of the forceful messages about risk *creation*, the risk creators, the costs and consequences which appeared in the GAR report? What of the recognition that risk creation was complex and that addressing all its aspects required close linkages between disaster reduction, climate change and sustainable development (the latter two themes the subject of the SDG and Paris Agreement frameworks being negotiated in the same year)? How did these messages translate into the final framework, if at all? As you will have guessed already, they were, indeed, lost in translation. Scanning the summary document of the final agreement (though after all that work the American delegation simply refused to sign up to elements of it),[5] there is indeed a goal to:

> prevent new and reduce existing disaster risk through the implementation of integrated and inclusive economic, structural,

legal, social, health, cultural, educational, environmental, technological, political and institutional measures.[6]

But as one drills down into the detail, the targets and priorities for actioning this aspiration disappear, only surfacing in a 'guiding principle':

'Build Back Better' for preventing the creation of, and reducing existing disaster risk.[7]

Building back better, as we have seen, is concerned with reconstruction according to sound DRR principles, rather than preventing risk creation. In fact in examining the full framework the concepts of addressing risk creation and risk creators are conspicuously absent. Lost in translation.

Colleague Lucy Pearson remarked in a paper (written with Mark Pelling) on the blurring in the final framework of the sharply focused targets of the original draft:

Much pre-conference discussion on specific levels of risk and loss reduction was replaced by the cover-all term 'substantially' … language became progressively more ambiguous as negotiations continued, with caveats such as 'where appropriate' appearing in many action-orientated paragraphs … targets quickly became non-quantitative and baselines went from to be 'implemented' to 'encouraged.'[8]

Sendai and other companion frameworks

Not only was the focus on tackling risk creation 'eroded' from the finalized framework, the intention to forge linkages with sustainable development and climate change was edited out. The politics of this complex assemblage of actors stripped that ambition out of the final text. Pearson and Pelling took a similar view of the loss of this vision for *coherently* linking the three frameworks, echoing Zenaida Willison's experience of negotiations ten years previously:

Whilst many governments made statements in support of linking disasters and development in the regional consultations and preparatory committees, behind the closed doors of the negotiation room they were less keen to put this policy coherence into practice.

Most notably, a strong push back to aligning DRR and climate change came from the US, who argued for separate conversations between these agendas. And, it seems they got their way. The final hours of negotiations saw the removal of references to all other post-2015 Frameworks from the Priorities for Action section. This seems to have been driven by the desire of agencies to protect their siloed mandates and the fear that connecting climate change and disasters would lead to the contracting of developed countries to take more financial responsibility for disasters.[9]

The word 'coherence' – forming a unified whole – has been adopted by the UN and Civil Society to describe the linking of different frameworks in a complementary way to achieve an overall goal. The hoped-for coherence of the Sendai framework with the separate climate change and sustainable development had been lost in its final form, but those other frameworks are strongly focused on aspects of disasters, development and sustainability. In fact, ironically, they have much more to say about tackling risk creation than SFDRR.

The Paris Agreement on Climate Change is explicitly concerned with systemic change to achieve its goal to 'substantially reduce global greenhouse gas emissions to hold global temperature increase to well below 2°C above pre-industrial levels and pursue efforts to limit it to 1.5°C above pre-industrial levels', recognizing that this would significantly reduce the risks and impacts of climate change.[10]

The idea that 'loss and damage' should be addressed by financial support from wealthy nations had been negotiated onto the sidelines of the Sendai framework and has also been a thorny issue in the continuing negotiations of the Paris Agreement. It was finally agreed in 2023, but with a small fraction of the target funding allocated and with the World Bank, seen by a spokesperson for small island states as 'pure gangster' and heavily dominated by US interests, as host to the fund.[11]

The *Sustainable Development Goals* include specifics related to energy consumption, manufacturing, climate change and environmental exploitation (though as with the SFDRR targets they are not quantitative but blurred by the word 'substantially'): including increasing use of renewables, decoupling economic growth from environmental damage, making industry cleaner, making cities safer, reducing subsidies for fossil fuel consumption, tackling climate change, protecting ocean and forest ecosystems.[12]

It seems that the concerns of the UNDRR 2015 GAR, under the headline *New risks have been generated and accumulated faster than existing risks have been reduced* are much more 'substantially' addressed in the companion frameworks – the Paris Agreement and the Sustainable Development Goals – than in the Sendai Framework for Disaster Risk Reduction. Because of this overlap we will continue to refer to all three in order to navigate through its investigative journey on the roots of hazard creation and magnification. In Chapter 10 we will discover the results of mid-term reviews of these frameworks which took place in 2023.

We've seen already that there are clear trends in emergence and magnification of hazards, and the GAR report we've considered highlights 'the inadequate pricing of disaster risk and of broader externalities in economic activity'. This statement, allied to Stiglitz's idea that 'There's a system where we socialize losses and privatize gains', suggests to me strongly that hazard creation and magnification are related to the dominant economic system which is structured around private enterprise, innovation and industry organized through pricing mechanisms which are able to evade pricing in disaster risks and other externalities. This, in a way, is a more pervasive system of 'global governance' than the structures of the UN. It's a system with a long history and which the GAR authors associate with growing human impact on the world and society in their section titled 'Welcome to the Anthropocene'. They assert that risk creation is outpacing the efforts of disaster risk reduction. How can we understand what is driving this trend, and why it is becoming increasingly problematic?

It seems that a key to this understanding is the idea of the 'Anthropocene' – the era in which human activity is having an increasingly significant impact on the state of our planet. In order to understand the roots of increasing risk we need to look historically at how this trend has emerged and intensified. Tracing the social, political and economic changes over this period, which we take as extending back at least 250 years, will reveal the ways in which risk is being created and magnified, with consequences for us all. This is the investigation we will undertake in Part 2.

Part 2

UNMANAGED GROWTH

Chapter 6

ADVENTURES INTO THE ANTHROPOCENE

What's the idea of the Anthropocene?
Humans, history and the planet

Why look back into history to understand current trends in creation and magnification of risk? My reasoning for this is that just as I chose to examine the events both before and after Hurricane Mitch, so understanding risk creation requires understanding the social, political and economic processes that drive this. Those changes date back at least to the social and political changes of the Enlightenment and gather form and pace in the industrial and consumer revolutions that follow. This section, therefore, will consider that history to discover how its trajectory leads towards an increasingly risky and unsustainable future. As we pursue this journey, I want to introduce three key players whose actions underlie it and who will be considered at milestones along the way:

- **Industry**: a term which I use to embrace not only manufacturing entities but the whole scope of the production system, including markets, firms, corporations, the financial system, resource extraction, product innovation, production, branding and marketing.
- **The public**: individually both as passive consumers and as active citizens, together as organized civil society.
- **Government**: which in this account focuses on the role of governance, recognizing the constraints and challenges of governance within nation states and also the complex issues of international and global governance.

I've termed this an 'Anthropocene' journey because in recent years it is claimed that human life has become the dominant influence on the climate and environment. This new phase has been named the Anthropocene period, but of course humans have made their mark on their environment in ways which stretch back into prehistory.

In the Chauvet caves, near Vallon Pont d'Arc in France, striking, simple, bold cave paintings created between 37,000 and 33,500 years ago retain their stark and dramatic contrast, almost as if they had been created yesterday. A child's footprint in the dust, an inquisitive visitor to the cave maybe, is thought to date to as much as 6,000 years after the art was created. Soon after that, 29,000 years ago, the caves were sealed by a rockfall and remained untouched until their rediscovery in 1994. Learning from the damage caused to the Lascaux cave paintings, these newly discovered caves have remained closed to the public, but a gigantic replica has been created a couple of kilometres away, and roaming the carefully recreated caves and paintings it is easy to suspend disbelief.[1] Viewing these early marks of civilization is striking. Despite the huge gulf of time there is an eerie sense of the presence of these early humans, and of their goals and ambitions.

Humans have always made their mark on their world but until three centuries ago the pace of growth, change and environmental impact had been slow. The world's population didn't tip the one billion mark until the early 1800s.[2] Since then change has been revolutionary and dramatic. In just two centuries the global population has grown to over 8 billion (see Figure 8).[3]

This and other accelerating changes in technology, production, construction and environmental impact have led to the idea of the 'Anthropocene', a term first coined in 2000 to describe a new 'epoch in which humans and our societies have become a global geophysical force', in a paper asking 'Are humans now overwhelming the great forces of nature?'[4] It presented a battery of charts of different trends – ranging from energy consumption and population growth to (rather bizarrely) the number of McDonalds restaurants – depicting what has become known as the 'Great Acceleration', particularly over the period from 1950 onwards and most significantly focusing on climate change (see Figure 9).

Figure 9 shows two key trends in this acceleration, the steep increase in carbon emissions, rising steadily from the mid-1800s and then accelerating more steeply from 1950. It also shows the upward trend in atmospheric CO_2 – again accelerating in the period from 1950. (The difference in rate of change between emissions and atmospheric CO_2 levels doesn't necessarily reassure us as we will see that the effects of

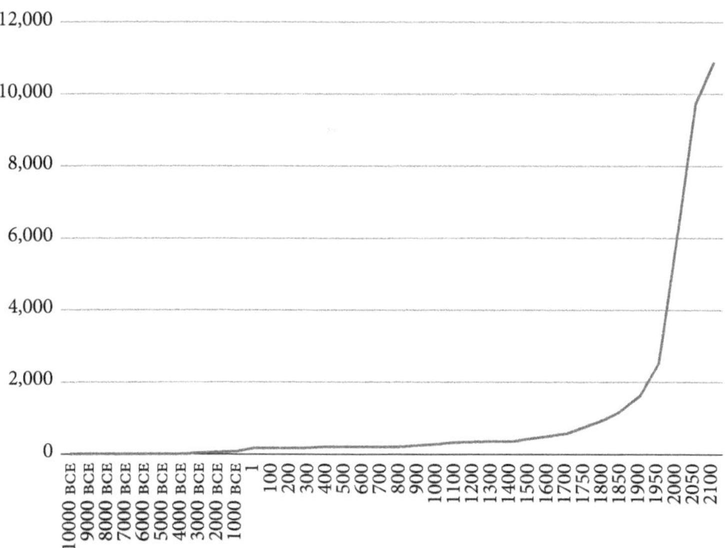

Figure 8 Population of world 10,000 BCE to 2100.
Source: UN-DESA/Statista. 2021 CC4.0.

the current rates of change and their forward projections may be substantial.)

The Anthropocene may be seen as commencing at around that time, 1950, but I suspect that for our purposes we need to look further back. Changes wrought during the period commencing with the Industrial Revolution are responsible for great progress in prosperity, health, longevity and technological progress (though to what extent you benefit from this depends on who and where you are).[5] Those changes may also be leading to creation and magnification of hazards, leading Beck to speak of the 'Risk Society'[6] and to recent concerted action on the negative effects of CO_2 emissions.

Even so, we must recognize that risk and disasters have always been with us. The famine of 1315–21, the greatest on record, devastated Europe. In the town of Ypres alone it wiped out one-tenth of the population between May and October 1316.[7] The Black Death (1346–53) is estimated to have killed 50 million people in Europe.[8] Earthquakes devastated entire cities, most notably striking Lisbon in 1755, when the city was almost completely destroyed and more than twenty thousand died,[9] leading to a sea change in thinking about the place of disasters in human experience.[10]

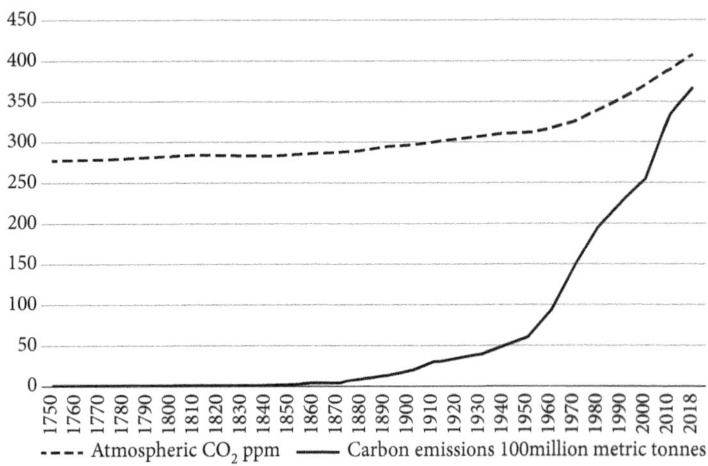

Figure 9 Trends in global atmospheric CO_2 and carbon emissions 1750–2018. *Sources*: icos-cp.eu/Statista. 2021 CC4.0; 'Atmospheric CO_2 Ppm by Year 1959–2022', Statista, accessed 8 November 2023, https://www.statista.com/statistics/1091926/atmospheric-concentration-of-co2-historic/; 'Global CO_2 Emissions by Year 1940–2022', Statista, accessed 8 November 2023, https://www.statista.com/statistics/276629/global-co2-emissions/; 'Global CO_2 Emissions by Year 1940–2022'.

The *hazards* lying behind these disasters were not generally created or magnified by human action, though early urbanization was a strong hazard magnifier of the Great Plague of London in 1665, of the Great Fire which struck the city in the following year and of the impact of the Lisbon earthquake, for example. The overall trend in *vulnerability* of the population over past centuries is hard to assess. One UK study tracked GDP per capita – which as a measure of access to resources has an inverse relationship to vulnerability – over six centuries, from 1270 to 1870, finding that this inflation-corrected measure more or less doubled over that period,[11] suggesting that vulnerability was trending slightly *downwards*.

Looking even further back we can trace early marks of human action. Already in those French caves there is evidence that the artists and later visitors used pitch torches to illuminate the caves. They possessed the secret of fire.[12] The shift to settled agriculture made use of fire to clear forests for farming and some argue that forest clearance for agriculture around 8,000 years ago and irrigation of rice 5,000 years ago led to increases in atmospheric CO_2 and methane, preventing the onset of an ice age.[13] Coal was used as fuel in China in the eleventh century, and in England from the thirteenth century. In fact, a UK commission to examine the evils of coal smoke began work in 1285.[14] Nevertheless, overall global trends in population growth and in carbon emissions showed modest increases over the whole period up to the nineteenth century.

Not one Anthropocene but three …

The Anthropocene marks a step change in the rate of change of human impact on society and its environment. As we seek the causes of increased risk, in particular driven by increased hazards, the revolutionary increase in human impact on the planet seems an important driver of those changes. We have seen already the sharp increase in population from the nineteenth century onwards, and also the steep acceleration of carbon emissions in that period. Examining this period may yield answers to questions about how risk is created and by whom. But when *exactly* did the Anthropocene start?

Some suggest a starting point at the onset of the Industrial Revolution, initially in England, marked by technological innovations such as Watt's development of the steam engine and by concentration of labour into factories using newly developed machines. Others set it much more recently, looking for a geological marker of human impact. It seems possible that a 'golden spike' in the geological record will be defined by sedimentary deposits showing the effect of post-war nuclear tests, detected in the mud of Crawford lake in Canada. Others question this specifically geological definition and argue for an Anthropocene related to social change.[15]

The important thing from our point of view is that the period, whenever it is defined, enables us to identify the processes of risk creation we are hunting for. Within the range which has been considered I have chosen to focus on *three* Anthropocenes in the hope of unearthing common themes which help us in our quest:

Anthropocene 1 (1776–1907) is a very English Anthropocene. Though encompassing the entire UK the focus of innovation, industrialization and urbanization was particularly on the North-West of England. This period starts, as others have suggested, with the refinement of the steam engine by Watt and Boulton in 1776.[16] While this serves as a marker at the start of the period, other key elements were the philosophical shift created by the Enlightenment and the ideological movement towards the free-market economy allied to the action of the 'invisible hand' marked by the publication of *The Wealth of Nations* by Adam Smith. Set free from earlier moral and religious frameworks for action, believing instead that markets would self-regulate for maximum efficiency, these changes in thinking and ideology triggered a wave of entrepreneurial, capitalist and innovative growth. The revolution was localized to England for nearly a century. Within this period, I aim to identify both positive and negative trends in hazard creation. I also aim to identify the drivers for these trends.

Anthropocene 2 (1908–49) shifts focus to America and is defined by 'Fordism'. The name suggests the assembly line production of cheap ('you can have it any colour you like as long as it's black') automobiles, but it has a wider meaning, encompassing the development of a prosperous workforce who become the customers for these and other products, thereby defining a *consumer cycle* and a constantly growing economy.[17] This Anthropocene spans the entire North American continent, on a much larger scale than the English Anthropocene, and again I seek to identify the positive and negative trends in hazard creation and find what or who lies behind these, and in particular the influence of this period on what came next …

Anthropocene 3 (1950–) corresponds to the likely scientific definition of the Anthropocene, and also marks the period of the 'Great Acceleration' in markers of human progress and impact on the planet. Whereas Anthropocene 1 was national and 2 was continental, Anthropocene 3 is global.[18] The waves of industrialization spreading outwards from England, Europe and America now reached the post-war economy of Japan, the tiger economies of South-East Asia and the expanding economies of the BRICS[19] countries.[20] Nations interlocked commercially, financially and also environmentally. The scale of production and consumption started to push at planetary boundaries.[21] What are the positive and negative trends in hazard creation? How and by whom are they being driven?

Ancient ancestors left images on cave walls, marks of their existence. Those arguing for the Anthropocene era suggest we are leaving much larger and sometimes brutal marks on our planet. This journey into the Anthropocene seeks to identify the benefits of global transformations and also the underlying drivers and mechanisms of risk creation.

Finally it looks into the future, examining modelling of future trends and impacts on society and its world. If, as UNDRR's reports claim, risk is being created faster than risk reduction can reduce its effects then we need to learn from this Anthropocene history in order to identify options for managing the future, seeking to leave marks of which we would be proud.

Chapter 7

ANTHROPOCENE 1: A VERY ENGLISH ANTHROPOCENE

Lit by the fiery glow of red-hot liquid iron poured from the furnaces, workers run with long-handled ladles filled with molten metal from the flames to steel-framed sand moulds on the floor of the factory. In lieu of proper safety wear they bandage thick card to their arms to protect them from splashes of the molten metal. Smoke and black dust fill the air. The cooling castings are knocked out of the moulds and fresh moulds pressed into wet sand, the castings stacked, still radiating heat, to be ground and finished.

Away from the factory floor at meal breaks workers, all male, shuffle patiently in long queues, holding battered tin plates, to collect food from a serving hatch. Once the working day is ended, they retire to dormitories where they sleep four to a room. And day by day the cycle is repeated.

This is not the Victorian England of the Industrial Revolution. This factory dates back not two centuries, but two decades. It's located in Shenzhen, China's third city, in Guangdong province, bordering Hong Kong.[1] It provides an impression, rather like the replica of the Chauvet caves, of the original revolution, which continues to reproduce itself as it ripples through the world, nearly three hundred years after its first iteration in the industrial cities of England. The UK long ago shrugged off much of its low-wage manufacturing industry, preferring instead to import goods from factories such as this (though today's zero-hour contract service industry has some parallels). Representatives of a global firm importing goods from Shenzhen factories witnessed these scenes. I was present as a film-maker recording their impressions and views. Some of the group expressed shock at what they witnessed, though their guide explained that the workers typically migrated from poorer rural regions of China and after several years accumulating their earnings could return home, buy land and establish themselves on a better economic footing than they could otherwise imagine. No-one complained, though in subsequent years there were for the first time reports of disturbances in factories as workers started to question conditions, for example, in the Foxconn factories, also in Shenzhen, where phones are assembled for Apple.[2]

There are striking parallels between the two geographically and historically separated localities. In both there was a transition from rural, scattered populations to concentration of labour in city factories. Health and safety took second place to the drive for production and profit. The underlying engine of industrial innovation and development drove economic growth. Complicit workforces were prepared to accept poor working conditions in return for a better income. In one respect the two situations are very different. The English industrial revolution was driven by markets, capitalism and innovation, by Adam Smith's 'Invisible hand'.[3] Chinese factories, and the 'Special Economic Zone' in which they are sited, are managed as part of the centrally controlled economic plan of the country. Nevertheless, the trajectory of development mirrors that of the earlier industrial revolution, progressing through a sequence from cheap basic products, towards more advanced and innovative manufacture, increasing in value.

Already, in this brief encounter with a latter-day industrial revolution, contrasting consequences in terms of prosperity and risk creation can be seen. In both cases industrial growth transformed countries from rural and agricultural to urbanized and industrially developed states. At the same time risk exposure through raw material extraction, manufacture, energy consumption and pollution grew dramatically, for example, through the massive increase in fossil fuel consumption and its effects. One study found that just as coal use decreased life expectancy in UK cities such as Manchester and Birmingham in the second half of the nineteenth century by about 9.8 per cent, more recently in the highly coal-polluted area north of China's Huai river, life expectancy reduced by 7.8 per cent.[4]

How did the industrial revolution achieve such prosperity and what do we learn about its role in risk creation? Who were the key drivers of change and who managed the costs? Governments? Industry? Wider society? Tracing the emergence of the first industrial revolution provides some answers to these questions.

Steam engines had been used since the early 1700s to pump water out of mines, but when these were succeeded by Watt's more efficient condensing engines they rapidly became the power source for new factories increasingly located in cities like Manchester – where labour was plentiful. Richard Arkwright established the first steam-powered cotton mill in Manchester in 1782 and many more mills followed. Manchester's population grew from seventeen thousand in 1760 to one hundred and eighty thousand in 1830 and by 1850 was manufacturing nearly 40 per cent of the world's cotton textile production.[5] The city skyscape was transformed by the hundreds of factory chimneys belching smoke day and night. The scene was welcomed by many as a sign of prosperity:

> Thank God, smoke is rising from the lofty chimneys of most of them. for I have not travelled thus far without learning, by many a painful illustration, that the absence of smoke from the factory-chimney indicates the quenching of the fire on many a domestic hearth, want of employment to many a willing labourer, and want of bread to many an honest family.
>
> (visitor to Lancashire, 1842)[6]

In some cases, factories wilfully concealed the hazards they were creating. An example is the use of white phosphorus in match production. Through a concerted campaign by the *Star* newspaper in the early 1890s it became clear that workers at Bryant and May's match factory were suffering terrible consequences of phosphorus exposure, what was called 'phossy jaw'. Their jaws would rot away and sufferers were at the very least terribly disfigured and in many cases died. It was found that the manufacturers had been deliberately concealing this, managing treatment of affected workers privately and ensuring physicians did not report the true cause of their illness. However, increasing media revelations and public pressure eventually led to a court case and minimal fines being imposed. It was several more years before the use of white phosphorus in manufacture was banned.[7] The health hazards of silicosis in coal mining[8] and of arsenic collected from tin mine chimneys in Cornwall by child labourers[9] were similarly minimized and even suppressed for the sake of profit.

Lying behind these technical and organizational innovations was the emergence of a free-market philosophy which had been formulated over several decades and found a totem in Adam Smith's

An Inquiry into the Nature and Causes of the Wealth of Nations (1776) with its idea of the balancing of supply and demand to reach an efficient equilibrium. The idea of an 'invisible hand' of free trade which achieved this without and even despite human intervention was seductive.[10] It built on the zeitgeist of the Enlightenment with its shift from a religious to a scientific perspective, liberating experimentation, problem-solving and innovation in search of profit, in an economic climate which welcomed it. Though Smith had previously published *The Theory of Moral Sentiments* (1759) and himself had a much more nuanced understanding of the role of commerce and of the morality underpinning it, this aspect of his thinking was largely unheard.[11]

In the face of seismic changes in trade, manufacture, urbanization and social organization determined solely by market forces and opportunistic innovation in this new free-market economy, management of the effects of change fell to government. It seemed that in the early years of the revolution it saw its main role as that of protecting the interests of the factory owners. Laws restricting trade unions had been in place for years and in 1799 and 1800 the Combination acts streamlined earlier laws, cutting out delays in trying cases so that any industrial action could be quickly closed down. In 1825 these laws were repealed, though it seems this was largely because it was believed they actually encouraged trade union action.[12] The government also acted to protect farmers' interests with the introduction of the Corn Law (1815), placing high tariffs on imported grain to protect domestic production, leading to a doubling of the price of grain and of bread.[13] It took until 1844 for the first health and safety legislation to be passed, covering cleanliness of factories and protection from moving machinery as well as extending limitations on child labour. This act was broadened from large textile mills to all workplaces of over fifty employees in the Factory Acts (Extension) Act of 1867, though the government's ability to inspect factories was tiny in relation to the number of factories.[14]

Beyond the business sector and government, the voice of the public, particularly of workers, was represented through media and campaigns, as well as being organized increasingly through unions. As working conditions were publicized there was increasing public outcry about child labour,[15] and in 1833 and 1844 laws were passed restricting the minimum age and maximum working hours of children in the factories.[16] Unions focused their attention on protecting wages and also on restricting entry to protect existing workers' interests. It seems they had limited interest in addressing the growing health and safety issues in the factories.[17]

As factories multiplied and their consumption of labour, raw materials and energy grew, their impact on the environment intensified. In 1848 naturalists first discovered a variant of the lightly coloured 'peppered moth', found on tree trunks in the newly industrialized cities, which was much darker in colour than the original variety. This biological change became a staple story of genetics, as it showed the mechanism of natural selection changing the population of moths from light to dark as a response to the blackening of tree trunks in the grimy cities, enabling the moths to maintain their camouflage. Much more recently geneticists have been able to isolate the mutation which produced the variant to around 1819, just forty years after factory chimneys started pumping out smoke.[18]

The impact of concentrated manufacture on Manchester was clearly visible. Residents reported smog restricting clear vision to a few hundred yards. By 1898 there were 1,200 factory chimneys in Manchester and another 760 in adjoining Salford. The smoke generated acid rain which eroded buildings and blackened everything. The rain and pollution were even killing plant life. Trees planted in new parks in the city had to be taken outside to the countryside to recover before being re-planted. Measures to reduce pollution were extremely limited, the main method being to build factory chimneys taller to spread the fumes further from their source.[19]

It was increasingly clear at the time that these conditions were impacting the health of the residents. When soldiers were recruited for the Boer War (1899), of eleven thousand who went forward, eight thousand were rejected as unfit and of those who joined up only twelve hundred were found moderately fit.[20] More recent studies have clearly linked pollution to ill health, concluding that it was a major cause of mortality, particularly in urban areas.[21]

It was over fifty years later, after an estimated twelve thousand premature deaths during the London 'Great Smog' of 1952, that a clean air act was finally established in 1956, though by this time some of the major causes of pollution had diminished with shifts in energy use and production methods.[22] Great Britain was no longer the 'Chimney of the World'[23] and the environmental impact of much of the massive uncontrolled growth of industry during the nineteenth century had been blown and washed away, leaving scars on the landscape as the record of extraction of coal and ore and the huge industrial landscapes of that era.

Risk creation in Anthropocene 1

One of the earliest alerts about the consequences of industrialization on the environment and atmosphere came from Victorian author, philosopher and art critic John Ruskin. From his rural vantage point in the English Lake district he observed the growth, industrialization and pollution of cities such as Manchester and in his lectures 'The Storm Cloud of the Nineteenth Century' referred to the 'Manchester devil's darkness' and the 'dense manufacturing mist'. His concern dug deeper than the simple physical consequences of industrial activity. He felt that society's abandonment of a 'moral ecology' was the root cause of the polluting storm clouds darkening the skies.[24] Certainly the impact of teeming factories in the expanding cities was dramatic. The polluted atmosphere, soil and rivers and the poor working, living and health conditions of its population were results of the 'externalities' of industry – costs which were not borne by the businesses and didn't have to be priced into their actions. Clear examples here are pollution and poor living conditions. Businesses didn't bear these costs, which fell instead on those working and living in the industrial cities. *Industry* therefore ignored these impacts of its activities, leaving protection of workers to others and pumping its pollution far and wide through higher and higher chimneys. Alongside neglect of externalities which created and magnified hazards, in some cases the protection of profit led to malicious management of them, as in the Bryant and May case and those of silicosis in coal mines and arsenic poisoning in tin mines. UNDRR's analysis of risk creation echoes this reality in present-day risk creation:

> The continuous mispricing of risk means that consequences are rarely attributed to the decisions that generate the risks.[25]

Regulation by *government* was often, at least initially, intended to protect the interests of industry. Regulations concerning working practices and health and safety were progressively, retrospectively introduced, but it was nearly a century after the onset of the industrial revolution that this basic health and safety legislation was extended to the majority of workplaces in the 1867 act. Even then enforcement was limited.

The *public* largely accepted the costs they bore, welcoming the benefits of employment and prosperity. Organized labour focused on protecting wages rather than health, and it was only through occasional media-driven campaigns, for example, against child labour and against use of phosphorus, that the public became drivers of change.

By the late nineteenth century the innovations which drove the revolution as well as its costs were being exported to Europe and America, eventually reaching China, but during that period England had been changed, reshaped and to some extent scarred by a very English Anthropocene.

Chapter 8

ANTHROPOCENE 2: 'ANY COLOUR YOU LIKE AS LONG AS IT'S BLACK'

If the skyline of Victorian Manchester was defined by chimneys, then that of early twentieth-century New York was defined by skyscrapers, from the eleven-storey 'Tower building' of 1888, to the dizzying heights of the Empire State Building at 1250 feet and 102 stories, completed in 1931. The heart of the city is Times Square. Today electronic hoardings plaster the high-rises above the streets, pumping out brand messages – Coca Cola, Apple, Ford, CNN. The square is a giant temple to the consumer dream. Its hoardings, graphics, brands and slogans symbolize a massive and entirely ephemeral industry – advertising. Ad agencies don't build cars, manufacture breakfast cereal or supply soft drinks, but they are and have been – since the skyscrapers were first being built – a vital part of the consumer cycle.

The first, English Anthropocene of the mills concentrated on basic goods – coal, cloth, steel. Now, crossing the water to the United States at the beginning of the twentieth century, a new wave of industrialization and production was starting, one which would be vitally dependent on advertising. And a key to this new transformation was 'Fordism'.[1]

Henry Ford built the first production line to manufacture the 'Ford Model T' and the first car rolled off the line in 1908, sparking a revolution. The production process broke down construction to a large number of steps along the line, de-skilling labour (and at the same time increasing its monotony). The cars were built of standardized and interchangeable parts so that different models could be built without customization. Through large-scale production of rigidly specified models (originally only available, despite the well-worn catchphrase, not in black but in grey, green and red), Ford's cars could be manufactured more rapidly and sold more cheaply than the competition and initially dominated the growing market.[2] The production line system paved the way for other complex consumer products such as washing machines and radios, but this was only one aspect of Fordism.

Cars needed customers. Over several years Ford raised the wages for his workers, who therefore became prosperous purchasers of his and other products (though the motivation for pay rises was high worker turnover rather than a deeper economic insight). Connecting customers with products required advertising and advertising needed brands to reassure customers of the quality of the products. Fordism, therefore, was a key element in an emerging and expanding 'consumer cycle'. At its simplest this cycle connected production and consumption. Production by industry was consumed by the public, generating revenue for the producers. This 'circular flow' was described in Knight's 1933 textbook *The Economic Organisation*[3] and Samuelson popularized it further using a plumbing metaphor to visualize the circular flow in his 1948 textbook *Economics, an Introductory Analysis*.[4] In fact, as can be seen in Figure 10, this cycle led to economic growth. The cycle was constantly expanding.

Despite the revolutionary impact of Ford's products, Henry Ford himself was not a great believer in the power of advertising to fuel the revolution. His early adverts were practical and factual, extolling the technical virtues of his vehicles, or simply pointing out that they were cheap:

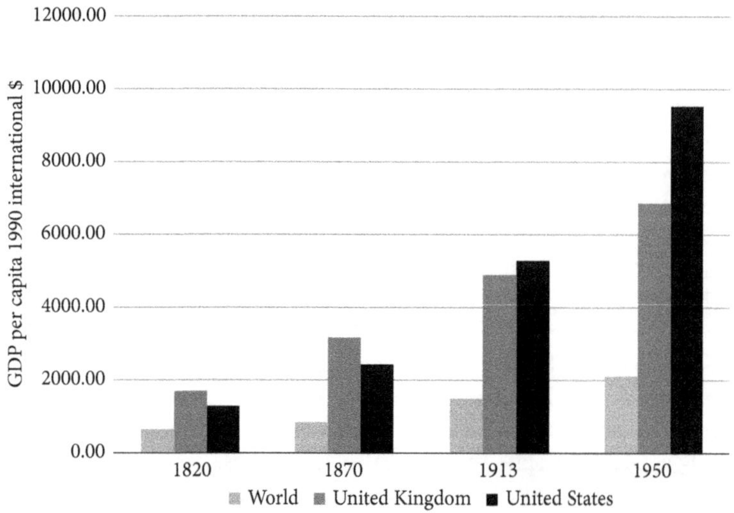

Figure 10 GDP per capita of world, UK and the United States.
Source: Based on data from Douglas Galbi. http://purplemotes.net. 2022.

The model T is a 4 cyl. 20 h.p. five passenger family car. Vanadium steel, the finest and costliest steel manufactured, is used throughout. Unit power plant with magneto as integral part of the same, – 4 cylinders in one block, water jacketed cylinder head removeable, offering easy access to all working parts of engine. (Life magazine 1 October 1908)[5]

Advertising was actually at the heart of the boom in consumerism, consumption and, as we will see, risk creation. By 1924 the Ford company had told engineers to stick to the day job rather than writing advertising copy. Ford's son Edsel realized even in those early days of the consumer revolution that products needed to be imbued with *desire* to fuel the boom.

Freedom, for the woman who owns a Ford. To own a Ford car is to be free to venture into new and untried places. It is to answer every challenge of Nature's charms, safely, surely and without fatigue. Where a narrow lane or a steep hill promises a surprise beyond, a Ford will take you there and back in comfort, trouble-free. Off and away in this obedient, ever-ready car, women may 'recharge the batteries' of tired bodies, newly inspired for the day's work. (October 1924)[6]

The car was no longer a machine constructed of Vanadium steel, but a lifestyle choice offering freedom, adventure into untried places and a recharge for the batteries. This was the job of advertising, to connect the growing disposable income of affluent Americans with the growing array of consumer products by making them objects of desire, status and identity rather than just functional machines, and in doing so to amplify the underlying ideas of autonomy and freedom as marks of consumer culture.

Over the period from 1900 to 1950, US advertising revenue grew from $450 million to $5,700 million, despite the shock waves of the stock market crash of 1929, the years of the Great Depression and the Second World War. It was a vital element in fuelling the ever-growing consumer cycle, leading US President Coolidge to say in 1926:

Advertising ministers to the spiritual side of trade. It is a great power that has been intrusted to your keeping which charges you with the high responsibility of inspiring and ennobling the commercial world. It is all part of the greater work of regeneration and redemption of mankind.[7]

Table 2 US Family Income, Non-necessities and Family Size

Year	Family expenditure	Non-necessities (%)	Family size
1901	$769	15.5	4.9
1950	$3,808	31.6	3

Source: Adapted from US Department of Labor. 2006.

During the period from 1900 to 1950 the consumer cycle boosted per capita GDP of the United States which accelerated ahead of the UK's (see Figure 10).

Growth in the prosperity of consumers in the United States was charted by the US Bureau of Labor.[8] Table 2 shows the growth of family expenditure for that period (though not corrected for inflation). The figures which I find interesting are the change in expenditure on non-necessities and the striking reduction in family size.

Non-necessities represent 'disposable income' shaped by choice and advertising. Where was this increase in disposable income, not spent on food, clothes and housing, directed? By 1929 car ownership had increased from an average of one for every three families to one per family. The boom triggered by Ford had given way to General Motors' programme of 'planned obsolescence' encouraging owners to upgrade their cars regularly and providing credit to do so. Alongside cars there was an explosion of other consumer goods purchased with the consumer's growing disposable income – radios, boilers, domestic appliances.[9]

Alongside the excitement of rapid growth in the 'roaring 20s' there were seeds of instability in the system. The stock market crash of 1929 and the Great Depression which followed led to manufacturing output falling by over 70 per cent and by 1933 unemployment in that country stood at 25 per cent.[10] Beyond the cities, mismanagement of farming methods across the Great Plains led to the environmental and social catastrophe of the 'Great Plains Dust Bowl Disaster'.[11] Heavy farming disrupted the fragile ecological balance of the region and drought and high winds in 1934, 1936 and 1939–40 tore the topsoil away, creating dramatic dust clouds. By 1940, 2.5 million people had left the resulting dust bowl.[12]

In the 1940s the economy turned around. The wartime manufacturing boom revitalized it and underpinned establishment of the suburban consumer America of the 1950s.[13] Internationally, the post-war economic stimuli created through by the newly established International Monetary Fund and the World Bank played a part in the export of this consumer-driven model of economic growth from the United States across the world.[14]

The footprint of the second Anthropocene

At the heart of this second Anthropocene story is the emergence of the consumer cycle. Society was transformed by the availability of consumer goods, the advertising industry that stimulated their desirability and purchase, the establishment of brands and brand identities to differentiate products, and the growth of consumer spending power among the prosperous sector of the population to enjoy this newfound affluence and the expanding range of products and services at their disposal. Progress was far from smooth, partly through the external circumstance of two world wars and partly through internal challenges, including the Great Depression and the Great Plains Dust Bowl disaster.

Dramatic industrial expansion and the growth of car use led progressively to 'externalities' including atmospheric pollution and environmental degradation. Compared with the UK the United States experienced less impact from these at first, stemming from the difference in population density. According to a recent estimate the UK has a population density of 280 per km^2 whereas that in the United States is 37 per km^2, in other words less than one-seventh of that of the UK. Therefore in the United States pollution and environmental damage were spread over a much larger area. Though the hugely lower population density of the United States mitigated the impact of industrial production, the stock market crash of 1929 and the Great Depression which followed led to massive unemployment and reduction in prosperity, stretching right through to the Second World War.

In the industrial cities pollution was accumulating, but it was not usually dealt with at source, but through 'end of pipe' methods, offloading it into the atmosphere, landfill or watercourses.[15] City smogs finally excited public concern and government action when in October 1948 in Donora, Pennsylvania, a thermal inversion blanketed the city in a smog produced by airborne industrial pollution which was so thick that all traffic movement was stopped because of lack of visibility. Twenty people died during the week of the smog. More persistent smog had been troubling the inhabitants of Los Angeles since the 1940s, attributed to industry and increasingly to motor vehicle emissions.[16] Though federal research budgets were authorized in the 1950s to investigate pollution the government dragged its feet, resisting legislation until finally the 'Clean Air Act' was passed in 1970.[17]

Nearly fifty years before this there had been an outcry over the particular issue of lead pollution. Automobile engine designers had been struggling to achieve smooth combustion and power, due to 'knocking', and it was discovered in 1922 that a lead compound added to fuel solved this problem. Immediately scientists raised concerns about public health as poisonous lead compounds would be pumped out into the atmosphere by the growing number of cars on the road.[18] The US Bureau of Mines was given the task of researching this, but was compromised in its independence as the study was funded by General Motors, which progressively restricted its freedom of action, firstly by imposing an embargo on any publicity during the research, and then by requiring them to refer to the trade name 'Ethyl' rather than Tetraethyl Lead Gasoline as they did not want the word 'Lead' to appear. General Motors insisted, further, that they be allowed to review the report before publication. A potential scandal erupted when several workers died from handling tetraethyl lead, but industry lobby groups blamed the workers rather than the compound, and also argued that its use was vital to industrial progress.[19] Despite increasing evidence of the cumulative effects of atmospheric lead exposure it continued to be used until nearly fifty years later the Clean Air Act had the effect of phasing it out, ushering in catalytic converters and lead-free fuel.

Wastelands created by the extractive activity of the second Anthropocene have also escaped 'pricing in' as costs of production. The US Government Accountability Office estimates that abandoned hard rock mines have contributed to contamination of 40 per cent of the country's rivers and 50 per cent of its lakes. The mining companies responsible for the contamination either disappeared or avoided liability for environmental impact of mining. The office gives the example of the long-abandoned Gold King mine in Colorado which, when disturbed a century after it was shut down, spilled three million gallons of contaminated water into the river system, contaminating water courses in three states. The clean-up of this one incident cost $63 million in financial settlements to affected communities. The most recent estimate, in 2021, was that the overall federal cost of clean-up of the abandoned mines runs at $613 billion.

Risk creation in the second Anthropocene

The economic transformation of the United States in the first half of the twentieth century was dramatic, transforming living standards and ushering in a consumer revolution. It was accompanied by growing hazard creation by *industry*, mitigated by the relatively low population density which meant that the environmental and social impacts of manufacture and energy consumption were generally limited. Industry generally avoided addressing impacts of production at source, relying on 'End of Pipe' methods, much as the English Anthropocene relied on taller chimneys. The legacy of hard rock mining echoes the idea of the generators of risk passing on the costs to society. The case of introducing tetraethyl lead into automobile fuel and the suppression of the dangers of this similarly echoes the case of phosphorus use in the UK. The Great Depression which spanned the 1930s introduced another poorly understood externality. The economics of growth, which seemed seductively plain-sailing during the roaring twenties, revealed poorly understood instabilities. The story of the period could be seen as one of both market failure and government failure, a story which has later echoes in the Asian market collapse of 1997 and the Great Depression of 2008–9.[20]

Government tended, as in the UK, to respond retrospectively to emerging social, environmental and economic impacts. The dust bowl disasters of the 1930s were reproduced in the 1950s, signalling a lack of effective management and regulation.[21] Cities experienced dramatic smog at more or less the same time as the Great Smog of London, but it took another two decades for government to move from research to establishing a Clean Air Act.

Over this time the *public* were largely acquiescent to the impacts of industrial activity, experiencing dramatic benefits from it and often remote from the effects of pollution and environmental degradation. The move towards eventual regulation was stimulated partly by the growing environmental problem of chronic smogs in Californian cities. The hardships of the Great Depression period led to some riots and campaigns but in general, particularly after Roosevelt introduced his New Deal, Americans appeared optimistic and trusting of their government.[22]

Exporting the American dream

The post-war era – the start of the third, global Anthropocene – was the point at which, with the restoration of peace, the American model of consumerism and economic growth started to be exported. This is the moment of the 'Great Acceleration'. With its low population density the United States had been able to absorb risk creation resulting from its activities but as its model of consumerism and growth went global, could the global impacts of risk creation be similarly absorbed?

Chapter 9

THE THIRD ANTHROPOCENE – MARCH OF THE MEGACITIES

We've pursued a journey from an English Anthropocene to an American, and reached a point after the turmoil of the Second World War where the consumer revolution is about to be exported from the United States. Cities have already played starring roles in the preceding sections: Manchester, New York and the industrial cities of the United States. The Global Anthropocene is the era of the city. The population was shifting from rural to urban and by 2007 more people lived in urban than in rural areas (see Figure 11).

The explosion of development after the Second World War poured into urban centres in Africa, Latin America and particularly across Asia. Populations grew, streets and buildings spread, city centre skyscrapers rose higher and higher, manufacturing expanded, shanties accumulated.

The soul of the city

Amid this economic expansion there is, and always has been, something deeper and significant about the city. Times Square might be a temple to commerce, where religious iconography is replaced by brand logos and worship is replaced by commerce, but it still strongly symbolizes the social change it represents. New York, in historical terms, is a very young city. Some are many older and similarly have strong symbolism underlying them. Kathmandu's population is under 1.5 million (2020),[1] tiny compared with megacities, those with populations over 10 million like Manila or Tokyo, but it's one of the fastest growing in the world. It's also a very religious city. Above the city's edge is the Swayambhunath stupa, one of the most sacred sites in Buddhism, and down in the city

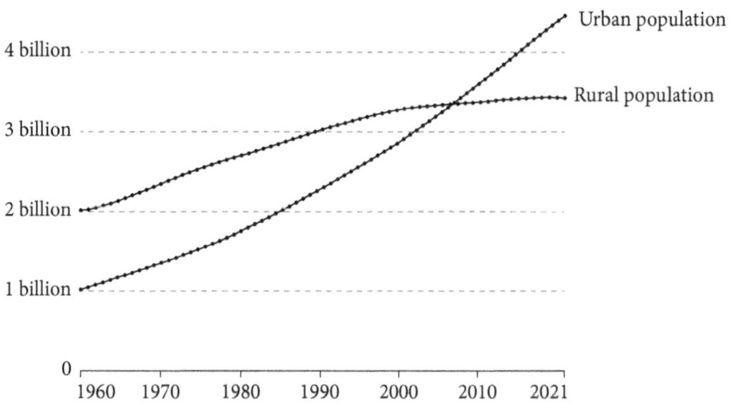

Data source: World Bank based on data from the UN Population Division
Note: Urban populations are defined based on the definition of urban areas by national statistical offices.

Figure 11 Number of people living in urban and rural areas.
Source: Our World in Data. 2023.

centre the Durbar squares are centres for both Buddhism and Hinduism. The Bagmati river running through the city is a holy site of temples and cremations. These facets of Kathmandu are reminders that cities are more than bricks and mortar. They are home, they are part of the identity of their inhabitants, and they represent movements and beliefs.

In post-enlightenment times different belief systems underpin the city. The commercial iconography of New York has spread to the expanding cities of the Anthropocene. One of its most striking icons is the Apple. Walk up Fifth Avenue from Times Square and you will encounter a glass cube, maybe fifteen meters on each side, standing in the centre of a wide paved area. Inside it is nothing. There's just a huge illuminated Apple logo on the front face. This is a cathedral to one of the dominant brands of the twenty-first century, dating back only five decades. Actually, there is more to this glass cube than meets the eye. Enter it and go downstairs and there you will find the Apple Store displaying its latest products. Go to the website to check when it's open and the answer is 'Always Open'.[2] Don't forget to book a shopping session with an online specialist. The parallels with attending worship, consulting or confessing with a priest at this cathedral of the brand are obvious. Just as missionaries attempted to spread their faiths across the world in previous centuries, so commerce and consumerism spread during the Anthropocene, and Apple is at the forefront, opening flagship stores with the same dramatic architecture in Seoul, Korea,[3] in Bangkok, Thailand,[4] and in Beijing, China.[5]

There is a serious point behind all of this. Cities need to be taken seriously. In many cases their expansion and development are random, unmanaged and sometimes chaotic, but as the homes of the majority of the world's population they matter. What they do to their inhabitants matters, and increasingly their footprint on the world matters too. The consumer cycle model exported from America and embodied in global brands such as Apple and Ford is only part of the story, but other aspects of that story, the individual cultures and belief systems of different localities are subsumed in many cases under that economic growth imperative of the consumer cycle.

Prosperous cities

What is the soul of today's cities? It's clear that a strong driver is economic. The lure of a better life drives inward migration and cities are now home to the majority of the world's population, they also host the majority of its productive capacity. They are the engines of growth and prosperity and the global trends are upwards. Over just twenty-five years (1997–2022), the global GDP increased from $31 trillion to $100 trillion[6] and global manufacturing output rose from $5.98 trillion to $16.33 trillion.[7] The World Bank's measure of global poverty showed a reduction of 10.6 per cent over the period from 2007 to 2019 in those living below the threshold of $2.15/day.[8] The 'Human Development Index', charting a range of measures of human development rather than just income, shows a steady upward trend, from 0.629 to 0.732 (2007–21).

Cities are the drivers of much of this growth in prosperity. According to a 2023 report, nine hundred major cities contribute 57 per cent of GDP – nearly $57 trillion. New York is the largest single city by GDP and outside America and Europe only Tokyo and Shanghai are in the top ten cities by this measure.[9] By 2040 however, the report claims that Beijing and Shenzhen will have joined this top ten, and that Asian cities will have increased their contribution to global GDP from 31 per cent to 41 per cent. Megacities like Manila will also grow dramatically with a GDP growth rate of 4.8 per cent per year. Across Africa and Latin America rates of economic growth will be lower, which combined with rates of population growth will increase levels of poverty amid their urban sprawl.

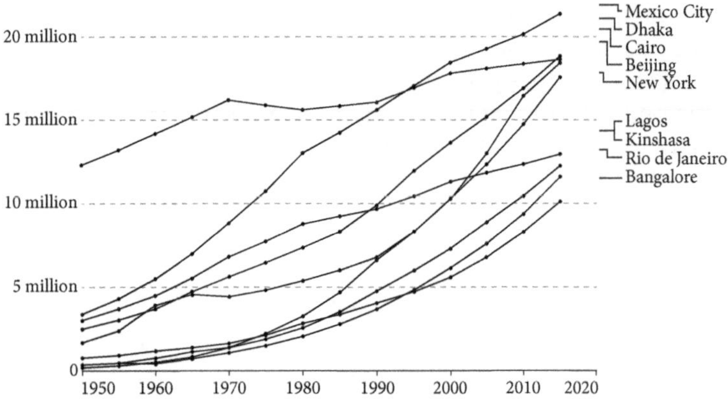

Data source: UN World Urbanization Prospects (2018)
Note: Data is available for the world's largest thirty cities (by population, in 2015). UN projections are based on its medium fertility population growth scenario and urbanization rates.

Figure 12 Populations of selected cities.
Source: Our World in Data. 2023.

In many cases, particularly in the period of rapid development and growth after the Second World War, city growth has been explosive and unmanaged. For example, Cairo, one of the world's megacities, grew in population from 2.5 million in 1950 to 21.75 million in 2022, multiplying its population by nearly nine times.[10] If you visit the Giza Pyramids today you have to be very careful to find a camera angle that doesn't include the city in the background as the urban sprawl has spread into the desert. Similar growth rates have been seen in other large cities across the world (see Figure 12).

A tale of three cities

In this Global Anthropocene the growth of cities, the stalling of rural population growth and the rural to urban migration trend – particularly in less developed economies – are striking phenomena. How do these trends play out in the cities of the world? What do these rising tides of increased productivity, prosperity and human development mean to city dwellers, and what are the costs of all of this in terms of risk creation? The following sections consider case studies of three cities: two – Nairobi in Kenya and Bangkok in Thailand – whose expansion has been unmanaged, driven by economic and population growth; and one unusually planned and managed city – Chandigarh in North India.

As we survey these cities you may find the account in some ways repetitive. Many cities suffer from overdevelopment, pollution, overdeveloped built environment, inward migration and social segregation – even the planned city of Chandigarh. Despite the contrasts between these cities, one in a less developed economy, one in a more developed and one which is intentionally created and managed, they face common challenges. Identifying the roots of these will help us understand what it is about this third phase of the Anthropocene which is driving risk creation and magnification.

Nairobi

Nairobi, Kenya's capital, had a population of 5.35 million in 2023, which had grown from just 137,000 in 1950, a thirty-nine-fold increase. It's estimated that by 2040 it will reach megacity status with a population of over 10 million.[11] In the centre, tower blocks cluster around highways, standing above big brand shops, arcades and malls. The city accounts for much of the nation's economic growth, generating 27.5 per cent of the entire country's GDP,[12] which in turn is trending upwards at an average rate of 5 per cent per annum.[13]

The intensity of traffic in this commercial hub creates high levels of air pollution. A rare social media posting during the 2020 coronavirus lockdown excited disbelief, showing Mount Kenya clearly visible behind a Nairobi city tower block.[14] While many maintained the photo was faked as they could *never* see the mountain, in fact that and sightings of Kilimanjaro from Nairobi's national park[15] were genuine, the result of the air temporarily clearing while traffic was off the roads during the lockdown. The harsh reality of urban pollution levels lies behind this story. Low visibility is an indicator of this,[16] and levels in Nairobi are way above WHO guidelines. Black carbon levels (produced by badly maintained diesel engines) are among the highest in the world.[17] The constant stream of motorbikes, scooters, matatas, cars and trucks drives much of this pollution, also accounting for 45 per cent of the city's total CO_2 emissions.[18] In an attempt to reduce congestion the city government has invested in road development, for example building the Nairobi expressway linking the city centre to the airport.[19] Travel times for car users have been dramatically reduced, but there are losers as well as winners – in this case building the road led to the demolition of thirteen thousand homes in the Mukuru Kwa Njenga slum.[20] The residents were clearly losers.

Nairobi's population growth is driven by the expansion of slums such as this around its periphery, resulting from rural to urban migration as well as natural population growth.[21] Kenya is a lower-middle-income country in world terms,[22] and studies suggest cities in this category tend towards greater segregation with lower-income inhabitants pushed towards slum areas, often towards the city's edges, away from the prosperous centre.[23] UN-Habitat defines slums as areas without access to improved water, access to improved sanitation, sufficient living area, durability of housing or security of tenure.[24] Nairobi's slums account for just 6 per cent of the city's land area but 60 per cent of its population.[25]

Kibera slum on the urban periphery of Nairobi, Kenya, is a typical example. Only about five square kilometres in area, it is home to an estimated two hundred and fifty thousand residents and is probably the largest urban slum in Africa.[26] A dense network of shacks has accumulated organically, bisected by the main railway line running through it on an embankment. The view from the tracks is of a carpet of rusted corrugated iron shack roofs, narrow dirt tracks weaving between them (Figure 13).[27]

The slum's expansion has been driven by population growth and inward migration from rural areas. Often men have come to the city in search of work and money, leaving families behind in rural areas.

Figure 13 Kibera Slum, Nairobi, Kenya. 2013.
Source: Michal Huniewicz. 2013. C.C.3.0.

Sanitation points are communal, as are water supplies. Many of the population live and work in the informal economy, working for cash or barter, often unregistered and paying no tax.

Slums like this accommodate over one billion people worldwide, and though the *percentage* of people living in slums is static, because of population growth the absolute *number* is still growing.[28] The density of housing and poor sanitation lead to greater disease spread and child mortality. Nevertheless, data for Nairobi slums shows that the standard of housing and services is higher than in rural areas, and nearly universal primary education is likely to contribute to better employment prospects and income for the next generation.[29] In addition a series of severe droughts in some rural areas, attributed by some to the effects of climate change[30] have led to increased hardship in rural areas, increasing the trend towards urban migration. Nairobi's shape and expansion is determined by social and economic pressures in the city and in rural areas – particularly the pull of rural populations to the perceived prosperity of the cities – much more than by any overall planning of its habitations and infrastructure.

Bangkok

Riding the Skytrain from the airport towards Bangkok's centre the view is of skyscrapers clamouring around and above each other (Figure 14). In the city's heart, railways, roads and walkways often double up above

Figure 14 Bangkok City Centre. 2010.
Source: Vyacheslav Argenberg/http://www.vascoplanet.com/ CC.4.0

the city streets. In front of sleek buildings are street food traders and market stalls heaped with global brands, many fake. Apple Central World is, however, the real deal, along with many other global brand stores.

The high-intensity lifestyle extends to eating out constantly. I discovered from a colleague based in Bangkok who came to stay and was intrigued by my cooking that apartments in that city don't usually have kitchens.[31] So eating out for many is a necessity rather than a luxury.

Contrasting with Nairobi, it's a rich city, forty-sixth in the global cities league table for GDP.[32] Its Human Development Index at 0.839 (2019) ranks as 'very high' compared with Nairobi's at 0.636 (medium).[33] The city's explosive growth has expanded its population from 1.36 million in 1950 to 11 million in 2023, taking it into the megacity league.

Known as the 'Venice of the East' because of the network of canals running through it, the waterways have a less romantic aspect as their banks are also home to many of the 20 per cent of the population living in slum conditions.[34] Similarly to Nairobi, much of this growing population are the result of inward migration, both from rural Thailand and those from other countries seeking work, particularly from Myanmar, Cambodia and Laos. Six hundred thousand foreign migrants are registered in the city and it is thought many more are unregistered. Migration fuels the 1 per cent per year annual growth rate.[35] Working in Thailand's rural north-east I met women who'd established decent homes there after a time working in the city. The story was similar to that of the Shenzhen factory workers coming from the country to raise capital, but in this case their work had been in the sex bars of Patpong street, part of the buzzing tourist scene. Many, like them, moving into the city live in the canal-side slums where they are exposed to disease as well as high risk of flooding, driven by the city's location on the Chao Phraya river which runs through the city to the sea.

Not only the canal bank dwellers but all the city's inhabitants are vulnerable to flooding, magnified by the huge proportion of the city under concrete, reducing the ground's absorptive capacity.[36] The risk exposure of the city and the region sometimes has far-reaching impacts. In 2011, floods striking much of the city also engulfed the Ayutthaya industrial area, stopping production at major companies including Honda, Toyota, Western Digital and Seagate. Nearly 50 per cent of global computer hard drive production was housed here and much of it was shut down.[37] The floods had a global impact on supply lines, creating global component and production shortages particularly in the car and computer industries, magnified by those industries' 'Just in Time' processes which meant they had little back-up inventory to draw on.[38] Overall, such events highlight the city's increasingly fragile resilience as continued development magnifies housing pressures and flood risk. In fact, it is estimated that the built environment of the city is sinking at a rate of up to two centimetres per year, increasing its overall exposure still further.[39]

Away from the canals, rivers of vehicles move constantly on and above the city streets, on ground level and elevated freeways. Traffic is intense. To a Westerner the city feels particularly hot, humid and often streets are filled with traffic fumes. Air pollution and smog in the city are so intense in dry seasons that people are often advised to wear masks and stay within air-conditioned buildings.[40] In one episode in 2023 a thick smog led to two hundred thousand people being hospitalized over one week.[41] Levels are usually four times the WHO recommended particulate levels and sometimes over eight times higher. Particulate levels in the city increase mortality significantly more than in many cities, associated with higher temperatures and sustained exposure.[42] Studies show that those working in the informal sector and living in slums have higher exposure to air pollution than other residents.[43] Despite these effects a large majority of the population prefer private transport to getting on the bus or the Skytrain; 81 per cent of commuters use cars and motorbikes in preference to public transport.[44] Transportation in the city is the single biggest source of CO_2 emissions, accounting for 38 per cent of the total.[45] The continued growth in road traffic has led the city government to build further roads, often elevated over existing routes like the Rama II expressway,[46] increasing still further emissions and pollution.

Bangkok's economic growth has been spectacular, though it is now slowing.[47] It is an exciting city to visit and travel round, but for its inhabitants its growth has magnified risk exposure, particularly through flooding and pollution. For poorer inhabitants, many pushed into slum areas, vulnerability is further magnified by poor living conditions affecting health and by increased exposure to seasonal flooding along the city canals.

Chandigarh

The development of both Nairobi and Bangkok is driven primarily by population growth, inward urban migration, economic growth and expansion, leading to increasing social and environmental challenges

Figure 15 Aerial view of Chandigarh. 2016.
Source: Muzammil/Wikimedia Commons CC 4.0.

resulting from this unmanaged development. Can a city be organized and managed for the benefit of its citizens? A city I've never visited consistently comes near the top of the league in media publications and travel guides as the best example of a planned and organized city.[48] Chandigarh in North India has an unusual history. Its foundation in 1950 came not only at the beginning of the post-war period but soon after the country's independence (1947). In the same way as New York was built on a grid system, the city's architects, including Le Corbusier, mapped it out on a 'superblock' system with a number of cellular neighbourhoods with traffic segregated. The city had well-developed infrastructure and services and building was restricted to low-rise structures (Figure 15).[49]

Since its establishment in the early 1950s it has grown dramatically, incorporating two adjoining cities into a metropolitan area with an overall population of over 1.6 million. It ranks third of Indian cities on the Human Development index (0.755)[50] and fourth in GDP per capita.[51] Studies find that it provides positive lives and livelihoods for the majority of its citizens.[52]

However, Chandigarh faces surprisingly similar challenges to those found in Nairobi and Bangkok.[53] Despite its foundation as a planned city, it has gradually accumulated slums around its periphery and despite decades of work to rehouse the occupants the problem has not gone away. Although over one hundred and seventy-five thousand people have been rehoused since 1981, slums still persist in the city,[54] and some studies suggest that city planning prioritizes the richer inhabitants.[55]

Like the other cases we have considered Chandigarh is also struggling with high levels of air pollution generated both within the city and in the surrounding areas. Particulate levels in the city are more than five times the WHO-recommended maximum and are trending upwards. While industrial emissions, particularly from brick kilns, account for a third of the total these are trending downwards; it's emissions from vehicles that are increasing. An additional factor is pollution resulting from burning off crop stubble in rural areas surrounding the city.[56] New road schemes are under way to address traffic congestion problems,[57] and in an attempt to reduce emissions a city-wide bike rental scheme is claimed to have reduced CO_2 emissions by 750 tonnes between 2020 and 2022. Chandigarh's story suggests that the challenges of inward migration, segregation of the poor into peripheral slums and ever-increasing vehicle use exacerbating other sources of emissions and pollution cannot be escaped even within this planned city.

Risky cities

Cities are the powerhouses of the global economy. Global consultancy firm McKinsey published 'Urban World: Mapping the Economic Power of Cities' shortly after the world's urban population passed 50 per cent in 2007. The report argues that 'The expansion of cities has the potential for further growth and poverty reduction across many emerging markets. Urbanization will be one of this century's biggest drivers of global economic growth.' Post-war growth spread across Asian cities with the 'economic miracle' of Japan followed by the emergence of the four Asian 'Tiger Economies' – Hong Kong, Singapore, South Korea and Taiwan. Others such as Thailand and Vietnam followed in their wake and while the economy of India has also been growing all these examples are dwarfed by the massive scale of Chinese productivity growth from the 1980s onwards. The magnetism of cities can be partly explained by their employment opportunities driving inward migration, partly by their access to diverse shopping, entertainment and leisure activities. Beyond that the city has a magic, an emotional lure. I talked to a father in a rural village in the Rift Valley in Ethiopia who was anxious for his daughter to get an education enabling her to escape the village for the city 'so she can escape this monotonous life that we lead'.

Associated with the mushrooming economic activity of city are magnified hazards – those environmental factors which can expose people vulnerable to them to disasters. Clearly flooding of the Chao Phraya river which runs through Bangkok is a hazard. The urban development of Bangkok has magnified the flooding hazard as the built environment has reduced the carrying capacity of the land, and indeed is leading to the city actually sinking, increasing its exposure to floods. Urban transport is a major source of emissions, and this is found to be generally true of megacities other than those sufficiently wealthy (and with political will) to force pollution control measures. One study of 167 cities found that transport and stationary energy were the two main sources of emissions and in a third of the cities studied transport accounted for more than 30 per cent of emissions.[58] In many cities, like those we've considered, road infrastructure expansion is a response to congestion, but this tends to create 'induced demand', stimulating even greater road use.[59]

One inescapable aspect of the growth of cities is, particularly in poorer countries, the segregation of their populations into prosperous and poor, the latter usually living in slum areas. While in some cases intensive disaster events impact people, for example, the 2011 floods in Bangkok and the 2015 Gorka earthquake in Kathmandu, these are infrequent compared with everyday disasters slum dwellers experience. Studies on urban risk in a range of locations highlight what are called variously everyday hazards, everyday risk and everyday disasters, agreeing that these account for far greater impact and losses than intense disaster events.[60] In the case studies we've considered, the poor and prosperous are segregated, the poor often living in slum areas like Kibera, often around the margins of more prosperous areas. A study of twenty-four global cities[61] found a similar pattern.[62] The shape of cities is driven by the distinct groups within them. While all cities surveyed had a tendency to segregate richer and poorer sections of the population this is more marked in poorer cities. In lower-income economies the richer socio-economic groups, such as the professional population, tended to gravitate towards the centre of cities, and sometimes to attractive coastal regions. The poorer socio-economic groups, low-wage workers and those right outside the formal economy are pushed to the edges of the urban region. Bangkok, a richer city, has an estimated slum population of 20 per cent. In Nairobi, a poorer city, it is 60 per cent and even planned and managed Chandigarh struggles to contain its growing slums. According to economic historian Brad de Long this segregation mirrors a general characteristic of growing global prosperity – if you are part of the economic cycle of earning and spending then the city will cater for you as a consumer, as a necessary part of the system, but if you are out of that cycle you are effectively invisible.[63] Slum inhabitants, like those in Kibera, contend with poor sanitation, cramped and unsafe living conditions, limited amenities and poor employment prospects, hoping that with improving access to education at least their children may be able to escape to become visible members of the city. Segregation in cities, between the poor and the better-off, demarcates vulnerability.

I have distinguished between the risk-creating results of hazard creation and magnification and the effects of vulnerability. The exploration of growing cities reveals that in these contexts both hazards resulting from overdevelopment and vulnerability resulting from inequitable and segregated development play a part in the risk creation process.

Accumulating urban risk

The urban cases described above, combined with studies of urban risk in a range of cities across Asia, Africa, Latin America, North America and Europe,[64] identify between them a wide range of sources of increasing risk, which are drawn together in Table 3.

Intensification of the built environment, poor planning and management of land use, growing industrial activity and social factors – particularly the consequences of inward migration and poor integration – all contribute to risk creation, leading to magnified hazards and vulnerabilities. Note that several of these factors have global impacts too. These in turn reflect back to the cities, for example, emissions leading to global warming which in turn magnify heat island effects in cities.

Risk creators in the city? Industry, government and citizens

Risk creation processes in the city are driven by the combination of industrial activity leading to economic growth, in turn leading to continued development and also attracting inward migration. The cases outlined above and broader studies both show that planning and management of urban growth are often very weak, and in its absence this wide range of hazards and vulnerabilities are created and magnified, resulting in wider global impacts as cities grow.

Table 3 Risk Creation in Cities

RISK CREATION IN CITIES	
Cities 1760–2025: Risk Creation	**Hazards and Vulnerabilities**
BUILT ENVIRONMENT	
Poor services, water, waste, sanitation, etc.	Magnifying health vulnerabilities
Informal housing: slums and shanties	Magnifying health and psychological vulnerabilities
Overdeveloped built environment	Magnifying climatic hazards, i.e. flooding
Increasing road transport, infrastructure, pollution and emissions	Magnifying local health hazards and climate hazards (also global impacts)
Lack of public transport and safe walking and cycle routes	Increased vehicle use, emissions and pollution hazards (also global impacts)
Hazard-prone structures	Magnifying climate and geophysical hazards
LAND USE	
Urban and suburban sprawl	Magnifying climate and health hazards through increased vehicle use (also global impacts)
Development on low-lying and unstable areas	Magnifying climate hazards, i.e. flooding
Degrading natural environment and biodiversity	Magnifying ecological hazards (also global impacts)
Lack of green space and tree cover	Increased physical and mental health hazards, decreased air quality
INDUSTRIAL ACTIVITY	
Exploitation of workers: low wages, insecurity	Magnifying economic vulnerability
Hazardous working conditions	Magnifying health vulnerability
Increasing materials consumption	Magnifying environmental hazards (also global impacts)
Increasing pollution	Magnifying health and environmental hazards (also global effects)
Increasing emissions	Magnifying health and climatic hazards (also global impacts)
SOCIAL IMPACT	
Inward migration from rural areas	Growth of health hazards in economically segregated informal housing
Lack of access to health services	Increased health and disease vulnerabilities
Lack of access to education	Persistent poverty, crime
Increasing segregation into poor/prosperous	Magnifying economic vulnerability
Social isolation	Magnifying psychosocial vulnerability
Weak links between CSO and local government	Inability to match services and risk reduction to local knowledge and capacities, magnifying hazards and vulnerabilities

At the outset I suggested that cities have or should have a soul. They are important as homes to the majority of the world's population. But these homes and their quality of life are being undermined, leading to a diminishing quality of life as the environment of the city is degraded, health is impacted, social capital is eroded (noted in several of the studies) and communal capacity to claim rights for better services is diminished.

The consequences of risk creation in cities are global as well as local. CO_2 emissions from cities account for over 60 per cent of the global total. Environmental exploitation and degradation by industry, including agro-industry, reflect back on the life of cities through increased heat and increasing frequency of extreme weather events, for example.

Cities and the planet

Rural population growth has stalled and is in slight decline (2022), but still nearly 3.5 billion people live beyond the city limits.[65] More than 75 per cent of the world's poor live in rural areas.[66] A detailed study of human migration patterns, drilling down to national and sub-national level, shows a clear global trend of migration from rural to urban areas which is particularly pronounced in low-income countries, suggesting that high levels of rural poverty drive people to the cities in search of a better life. (The opposite is true in high-income countries where in some cases there is net urban to rural migration as people escape the cities for a better life.)[67]

Projections from 2020 to 2030 suggest that rural poverty will decrease over that period, though this is partly a consequence of rural to urban migration and partly that of overall economic growth over that period. Meanwhile the same study suggests there will be no reduction in urban poverty levels over that time.[68]

Rural areas grow the food that feeds the cities. Though in Nepal, Kenya, India, Thailand and many other countries rural poverty is driving people away from the land to the cities, growing agricultural production is vital. In 1968 a USAID administrator, William Gaud, coined the term 'Green Revolution'. He identified development of new seed varieties, fertilizer use, new farming methods and government policies as elements of this revolution.[69] Since he spoke these words total agricultural output has grown from $1.34 trillion to $4.2 trillion (2019).

But the need to grow more to feed a growing population brings other consequences. About 26 per cent of total greenhouse gas emissions come from food production (cattle production producing between *20 and 60 times more greenhouse gases* than grain), and fertilizer and other nutrient run-off from agriculture accounts for 78 per cent of ocean and freshwater pollution. Growing agricultural production now utilizes 50 per cent of the world's habitable land.[70] Rural farmers, meanwhile, don't necessarily benefit from this green revolution, as in many cases small farmers can't access seeds, fertilizers and pesticides as bigger producers can, and sometimes they are displaced in favour of those bigger producers through government policy or land grabs.[71] And so, faced with these pressures, migration by the poor from rural areas to the cities continues.

Cities and the planet: The Great Acceleration

The story of the third Anthropocene is largely defined by the development and expansion of cities, driven by pressures of natural population growth, inward migration, economically driven development in the rich centres, providing slick homes, shops and entertainment for their wealthier inhabitants while the poorer populations live in sprawling slums towards their margins, taking low-wage work or operating in the informal economy. In 2025 it is estimated that the six hundred largest cities will generate 60 per cent of global GDP.[72] Their growth in production and consumption has contributed to an 8,500 per cent global growth in GDP, from $1.3 trillion in 1960 to $85.2 trillion in 2020.[73] Cities are also responsible for over 60 per cent of global carbon emissions, transport and buildings being among the largest contributors.[74] Standing at any road junction in Kathmandu is enough to remind you that we are, as American President George W. Bush said, 'addicted to oil'. Fossil fuel consumption drives escalation in CO_2 production – up from 6 billion metric tonnes in 1950 to 37.49 in 2022, an increase of over 600 per cent.[75] These and other accelerating trends over this period have led some to describe it as that of the 'Great Acceleration'.

The term was introduced in 2007[76] to highlight accelerating post-Second World War growth in a number of dimensions of global economies and their impacts on world systems. The most critical of these are population growth, production growth measured as GDP, urbanization and energy consumption.[77] While there is evidence that some trends are slowing, the key drivers of change, population, production and energy consumption are not forecast to flatten in the near term, and these give rise in particular to the analysis and policy recommendations underlying the United Nations treaties we considered earlier, particularly the Paris Treaty on climate change.[78] Taken at face value they present us with extremely disturbing scenarios for the future of the planet and its people. The same trends which have brought the prosperity many enjoy may be transforming, according to some assessments, into a global nightmare as the exploitation of the planet expands beyond its carrying capacity.

*Running out of road? No chimney tall
enough, no pipe long enough?*

The Anthropocene trajectory, from polluted cities of the first industrial revolution, to the consumer expansion in the United States, and to its post-war global reproduction across the globe may, as you recall from the 2015 UN GAR report, be taking us beyond planetary limits – no chimneys tall enough, no pipes long enough to dispose of the by-products of expanding consumption of materials and energy. Ecological economist Herman Daly argues that we have moved from a world of 'empty-world economics' to a world of 'full world economics'.[79] Trends in greenhouse emissions and the impact on our breathing planet

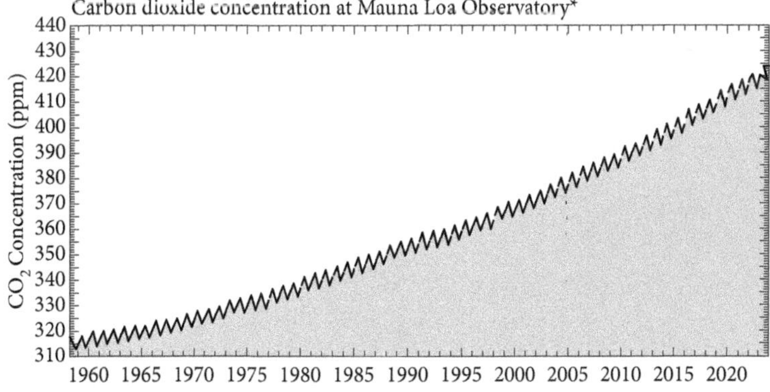

Figure 16 Carbon dioxide concentration trend.
Courtesy Scripps Institution of Oceanography at UC San Diego. C.C. 4.0.

are just one example of ways we are pushing beyond planetary limits[80] … running out of road.

The 'Keeling Curve' is well known as one of the earliest demonstrations of the accumulation of atmospheric CO_2, plotted continuously since 1960.[81] Dave Keeling found that over each year CO_2 reached a maximum in May and then fell to a minimum in October. He said 'We were witnessing for the first time nature's withdrawing CO_2 from the air for plant growth during summer and returning it each succeeding winter'. He was detecting the natural earth *breathing*. The graph gives an eerie sense of the living planet. But of course its breathing is running into difficulties. Year on year, the average level of CO_2 is rising (Figure 16).

The rates of growth and change we are experiencing in the Anthropocene are the marks of the Great Acceleration. Fossil fuel consumption resulting in increasing greenhouse gas emissions leading to risk creation through the consequences of global warming is just one of the trends of the great acceleration, but a major one. Climate change, other drivers of increasing unsustainability, and their consequences in risk creation rather than risk reduction are the foci of the three UN frameworks established in 2015, which all reached their midpoint reviews in 2023. What did these reviews find about progress in turning the world away from these trends in risk creation towards sustainability?

Chapter 10

HALFWAY TOWARDS DISASTER

Mid-point reviews of UN frameworks

When the 2015 UN frameworks for disaster risk reduction, sustainable development and climate change reduction were established, trends in growing disaster risk were already deeply concerning. The aspirational goals set by those frameworks envisioned a turn-around in global fortunes by their endpoints in 2030. As these frameworks were introduced I flagged up the fact that despite their global scope they have little power to exercise global governance. We saw, for example, how member states negotiating the SFDRR disasters treaty adapted and edited it to meet their national concerns. In this process, as with the others fashioning the sustainable development framework and the climate change treaty, we witnessed *international* rather than *global* governance. The distinction I made is that these processes attempt to blend together diverse national priorities into an international agreement, rather than having the ability to enforce legislation globally to manage the three programmes. There's no global government. I've laboured this point because it accounts for the gulf between the goals of these frameworks and the limited progress towards them. This reality was brought into sharp focus as the three frameworks underwent 'midterm reviews' in 2023. Compared with the hopes of the frameworks, the reviews made sobering reading (my emphases throughout). The SFDRR disasters framework review reported:

> '*despite commitments to build resilience, to tackle climate change, initiate just and equitable energy transitions, redress declining biodiversity, renovate food systems sustainability, address deep-rooted water resources issues, and pursue sustainable and regenerative development, current societal, political and economic choices are*

doing the reverse. <u>Intensive and extensive risks are growing at an unprecedented rate. Human actions continue to push the planet towards its existential and ecosystem limits, intensifying risk.</u>'[1]

It reflected further on the progress of all three frameworks: '*With growing uncertainties and increasingly complex risks, amplified by increasing disaster impacts and losses, <u>belief in our collective ability to achieve the 2030 Agenda appears to be waning.</u>*'[2]

The Sustainable Development Goals report was similarly bleak in its assessment:

'*It is time to sound the alarm. <u>At the midpoint on our way to 2030, the Sustainable Development Goals are in deep trouble</u>. An assessment of the around 140 targets for which trend data is available shows that about half of these targets are moderately or severely off track; and over 30 per cent have either seen no movement or regressed below the 2015 baseline.*'

'*Leave no one behind. That defining principle of the 2030 Agenda for Sustainable Development is a shared promise by every country to work together to secure the rights and well-being of everyone on a healthy, thriving planet. But <u>halfway to 2030, that promise is in peril. The Sustainable Development Goals are disappearing in the rear-view mirror,</u> as are the hope and rights of current and future generations.*'[3]

Conclusions of the Climate Change Agreement 'Global Stocktake' are buried in a mass of paragraphs couched in technical language. A separate 'emissions gap' report highlighted the gulf between the agreement's goals and reality:

Unconditional and conditional NDCs for 2030 are estimated to reduce global emissions by 2 per cent and 9 per cent respectively, compared with current policy projections and assuming they are fully implemented. To get to levels consistent with least-cost pathways limiting global warming to below 2°C and 1.5°C, global GHG emissions must be reduced by 28 per cent and 42 per cent respectively.[4]

Despite the bleak assessments of these reviews the sense of urgency expressed in them didn't drive a dramatic change of course. Let's look at a case study of how this plays out – the COP28 meeting (November–December 2023) responding to warning messages about lack of progress in managing climate change. This meeting drew huge media attention because though intended to address the challenges of the 'stocktake' the meeting was held at what climate activists would think of as enemy HQ: Dubai, in the United Arab Emirates, a major fossil fuel producing nation. The event was chaired by the head of the state oil company, Sultan Al-Jaber. What kind of deal could be brokered in such a setting? Negotiations over the final text continued over several days and nights. When the draft emerged, for some it was seen as a rare triumph as for the first time in COP28 meetings the issue of fossil fuels was finally put on the table. But for others, including small island states most affected by climate change, the text represented yet another weak compromise.

Whereas the stocktake's key finding had been the necessity for 'phasing out all unabated fossil fuels' the final wording blurred this into 'Transitioning away from fossil fuels in energy systems, in a just, orderly and equitable manner'.[5] This allowed for a huge range of interpretations of the timing and scale of 'transitioning', at odds with the sense of urgency of all the midterm reviews. Many felt that this compromise was driven by lobbying from oil producing countries and large energy corporations (a record 2,456 fossil fuel lobbyists were registered at the conference),[6] who continued to resist the urgency stated in all three frameworks, persisting on a road the midterm reviews suggested was halfway to disaster.

The midterm reviews served to emphasize the gulf between aspiration and reality in the three frameworks and in individual governments' support of them. The repeated blurring of language from concrete targets to 'substantial' change and from 'phasing out' to 'transitioning' sent a strong message that signatory governments lacked a sense of urgency. The hard-headed bean counters of the insurance industry see it very differently.

Your premium is increasing …

The concerns of the insurance industry are very different to those of the oil corporations. They need to know what premiums to charge you, and the whole world, to cover increasing losses. These companies are funded, in turn, by 'Reinsurance' companies, which underwrite their risk. The second biggest in the world is 'Swiss Re'.[7] It's in their interests to cut through the crap and make their best estimates of the consequences of our continued pushing beyond planetary limits. Their research study assesses the economic hit to global, regional and national economies by 2050 based on the IPCC's climate projections. To do this they've drawn on detailed data from Moody Analytics, one of the world's biggest risk assessment agencies. The point I'm underscoring is that this is a hard-headed business assessment.

They estimated that by 2050 the global economy would shrink by 10 per cent, and that some areas, particularly South-East Asia, would be hit particularly hard, shrinking their economies by 25 per cent. They described practical effects as cities get hotter, disease becomes more prevalent, coastal and low-lying areas (where many cities are situated) experience increasing floods, food shortages get greater and migration increases as some areas become un-liveable. They list eight 'impact factors' and show how these will affect everybody's lives in the world of 2050:[8]

- **Agricultural productivity:** Changing temperature and rainfall leading to decreasing crop yields
- **Human-health effects:** higher mortality and morbidity, particularly from spreading insect borne diseases
- **Labour productivity:** Reduced productivity due to heat effects, particularly in outdoor contexts such as agriculture
- **Sea level rise:** Land loss in coastal areas from floods, erosion and salination affecting the built environment and agriculture
- **Tourism flows:** Warm regions will become less hospitable, reducing tourism revenues for some destinations heavily dependent on it
- **Household energy:** Rising temperatures increase reliance on air-conditioning and related energy consumption, affecting energy prices and sources
- **Acute physical risks:** Increased likelihood of climate related disaster events
- **Migration, trade and biodiversity:** Unpredictable effects on social stability from increased climate driven migration, on trade from disruption of supply chains, and on ecosystems from decreased biodiversity

(Adapted from Swiss Re report 'The Economics of climate change: no action not an option' (2021)[9])

The report is driven by the organization's self-interest, assessing practical consequences of global warming as they have to insure these profitably, but as a result it brings into much sharper focus than many scientific communications how climate change will actually affect the public.[10] Many of the impacts it identifies are on life in cities, intensifying risk creation still further.

That's the possible world of 2050, but what about now? The UK *Financial Times* published an assessment of impacts of rising temperatures in cities, one of the impacts highlighted above. It charted those impacts right now, rather than in 2050. It drew on meteorological data to compare the number of days in which city temperatures were over 30°C in 2019–23, compared with temperatures in 1950–4. It found that there were 1.7 times more days over that threshold in Miami, 2.6 times as many in Rome, 5.7 times as many in Beijing, 8.1 times as many in Paris and 10.4 times as many in London.[11] The study also looked at the specific consequences of extreme heat in Arizona, charting those who had been recorded medically as dying as a direct consequence of it. Between 1970 and 1990 an average of sixteen deaths per year were recorded. In 2022 the number had risen to 257, reflecting a frightening upward trend over the whole period. The author of the *Financial Times* report says, 'I understand the focus on the 2C limit, but it a) sounds small, b) refers to some date in the future, c) lacks any connection to human experience and d) does a pretty bad job of describing what is happening with temperatures.'[12] Assessments such as those above show the practical consequences for peoples' lives and livelihoods, not only in the future but also in the present.

Potentially more threatening than these direct heat effects are the consequences on prosperity and food security, which according to the Swiss Re report are more severe in regions such as Asia and Africa than in Northern Europe and North America. The estimate of a downturn in South-East Asia's GDP of 25 per cent by 2050, for example, is dramatic and would in turn drive migration, social instability and conflict, adding to the other drivers of conflict in an increasingly war-torn world.

Incoherent futures?

I am struck by the dominance of climate change and the Paris Agreement in popular thinking, pushing the sustainable development goals and the disaster reduction frameworks into the shadows. I showed how negotiation of SFDRR gradually eroded links to sustainable development and climate change and noted that Pelling and Pearson suggested that a kind of inter-framework competition and territoriality was deepening the barriers between them. Whatever the specific drivers, the result is a lack of 'coherence' – joined up thinking.

The book *Disaster by Choice*, written by disaster studies expert Ilan Kelman,[13] emphasizes the many ways in which vulnerability and hazards are created and magnified, reminding us that despite the all-consuming focus on climate change this magnified hazard adds to a risk landscape that would be increasingly dangerous as a result of human choices, even without the effects of fossil fuel consumption.

The evidence of the midterm reviews of the UN disaster reduction, sustainable development and climate change frameworks is that choices which continue to be made create increasing risk across the whole planet. How and by whom are these choices being made? In Part 3 we investigate the drivers of risk creation and, accepting that international initiatives have limited ability to drive change, consider who and what might enable a change of course.

Part 3

SUSTAINABLE FUTURES

Chapter 11

IDENTIFYING THE DISASTER MAKERS

For centuries composers have written requiems – Mozart, Haydn, Berlioz and Faure, to name a few. They are musical settings for remembering the dead and the form has been used more widely to accompany losses from different causes, as with Britten's War Requiem. As I listen right now to the lesser-known composer Durufle's setting,[1] I am looking out on the fields and forest around where I live. My local environmentalist friend Ian is desperately sad about reduction of biodiversity even in this secluded rural landscape. As he roams the fields and woodlands, he can chart year by year the loss of bird and insect species. It's as if a requiem is playing across the landscape. A much more expansive requiem is playing over our urban landscapes as they sprawl out further, suck in migrants to poor quality settlements and cram roads and buildings ever closer together under urban smog.

The trajectory of risk creation – 'New risks have been generated and accumulated faster than existing risks have been reduced'[2] – as the UNDRR claims – is magnifying hazards and vulnerabilities in the growing cities of the world, and the effects of human-induced risk creation are felt everywhere.

Throughout the trajectory of the Anthropocene, we've seen increasing exploitation of raw materials, increasing manufacture, construction, building, consumption and pollution, leading to magnification of climate and geophysical hazards. We've seen the way that cities have been segregated in a kind of economic apartheid as the poor populations of cities, often migrants from even poorer rural settings, are pushed into slums and shanties on the peripheries while prosperous populations enjoy all the city centres have to offer. Even for those, their actions and choices are creating risk that affects them and the whole world. The quality of life for all city dwellers is diminishing.

As we've travelled through the Anthropocene the drive for progress and prosperity has fuelled the 'Great Acceleration'. The trends it describes – population, production and consumption growth, environmental exploitation, pollution, emissions, global warming – have reached and passed through planetary boundaries. Projecting these trends into the future predicts that a hotter, poorer, more unstable and dangerous world will be home for future generations. Composers are already at work creating requiems for our planet.

The structure of a requiem concludes, however, with the 'In Paradisum' – 'into Paradise'. In its religious usage it is played as the body of the person being remembered is carried away. In Durufle's composition it is an ending of peace and hope after the anger and sadness of the previous sections. As I listen it suggests a resolution, a tentative hope of something better, though I'm told the final note is, in musical terms, *unresolved*.

The goals of the United Nations developmental frameworks, such as those we've examined, seem written in this spirit. They express aspirations and visions of a better future – driving down disasters, managing development sustainably, reducing the impact of climate by successful mitigation and adaptation. Many would welcome these goals. As we've considered the growth and development of cities it's hard to imagine how these visions would translate into urban reality, but there's even a United Nations Framework for that. The 'New Urban Agenda', signed off in Quito, Ecuador, in 2016, pictures cities in which segregation is dissipated by access to services, decent housing, transport infrastructure and employment opportunities.[3] These future cities would be organized into compact localities providing work, services, green spaces and good amenities, reducing the need for travel to other areas of the city. Public transport would reduce the need for private vehicle use. Principles of risk reduction and of sustainability would be applied and the negative climate impacts of cities would be reduced.

Many would welcome the United Nations vision for future cities, as they would the aspirations for the future of the world. Unfortunately, as we have seen in the midterm assessments of three major frameworks, these visions are far from translating into reality. As the review of the Sendai framework put it, 'belief in our collective ability to achieve the 2030 Agenda appears to be waning'.[4]

This book is a search for answers to the questions Ai Weiwei set running in my mind when he referred to the 'disaster makers' responsible for the human cost of the Sichuan earthquake. Who or what are the drivers of increasing risk, frustrating the goals of United Nations frameworks and more importantly our hopes for a sustainable future? Based on the understanding we have developed of disasters, risk, vulnerability and hazards on our journey through an Anthropocene period stretching back over 260 years, the following sections set out to identify mechanisms of disaster making, leading to options for change towards the kind of sustainable 'In Paradisum' depicted in development frameworks. In doing so the analysis considers key actors who have been identified throughout the Anthropocene journey – *industry*, the *public* and *government*. I've repeatedly looked at the roles of these three key players during the Anthropocene journey. What I am primarily concerned about is key *functions* of these three as they affect risk creation, the state of the global commons and our safety and sustainability. To show you what I mean I've organized these 'players' as elements of a cycle.

The elements of the consumer cycle

Figure 17 relates these three elements in a cycle, production by *industry* leading to consumption by the *public*, and subject to management and regulation by *government*.

This is an extremely simplified cycle compared with those developed by economists. I mentioned earlier those devised by Knight (1933) and Samuelson (1948) and these have been elaborated by other economists, adding other actors, 'inputs' and 'leakages'. Raworth and others argue that these representations are of a closed system, continuing to neglect their relationship to resources, exploitation, pollution and environmental degradation. Raworth depicts the economy as part of an open system, the 'Embedded Economy'. This concept, part of what she calls 'Doughnut Economics', elaborates this idea in economic terms.[5] I am simply concerned here to consider how the three main actors in the cycle, however it is represented, affect risk creation.

Beyond city and planetary limits?

The roots of risk creation lie in the fact that this is not a steady-state cycle. The graphs of the Great Acceleration emphasize the effects of continual growth – in population, economic activity, production and consumption. A very simple way of visualizing this is through an expanding cycle. Figure 18 suggests this inexorable expansion. It symbolizes the consequences of the continual growth in the consumer

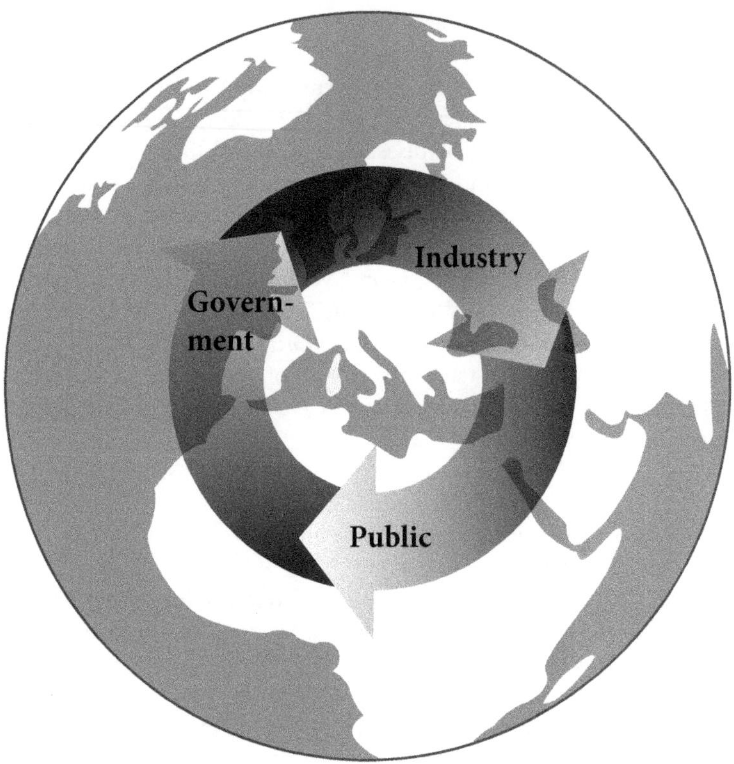

Figure 17 Consumer cycle.
Source: Author.

cycle for cities, leading to their growth in scale and accumulation of risk. It also highlights the consequences of this growth as we've moved to what Daly called 'full world economics',[6] and beyond even that to exceed planetary limits, as Rockström and colleagues showed in their influential 2009 paper 'Planetary Boundaries: Exploring the safe operating space for humanity'.[7] We've focused on cities, but they depend in turn on extractive industries prospecting, mining and drilling for the raw materials for manufacture, building, power and transport. Much of this activity takes place in low-income countries that see little of the economic benefit of businesses run from the rich North. Much of it also depends on exploiting low-wage labour. Until recently this huge aspect of the industrial system was largely hidden from scrutiny and many argue that recent moves towards transparency, for example, through

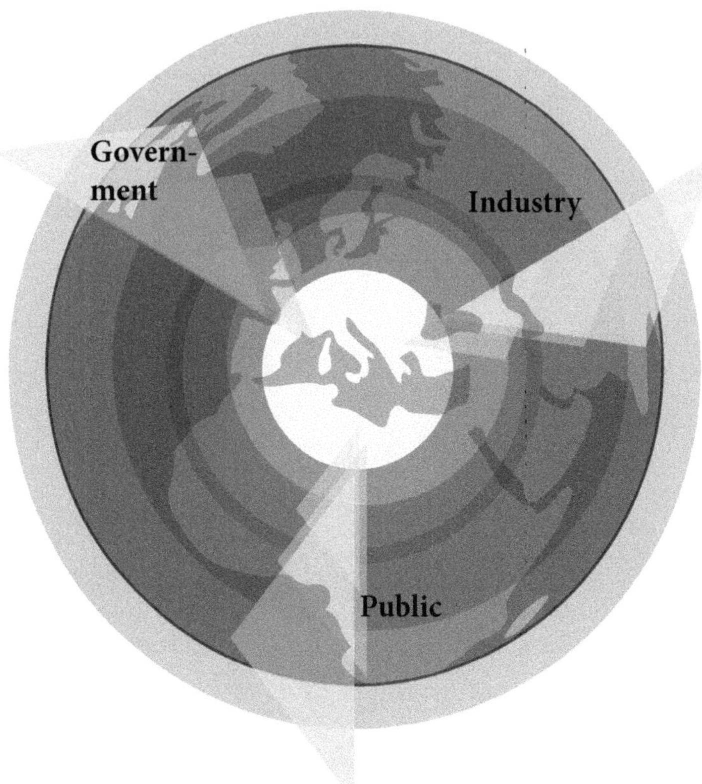

Figure 18 Consumer cycle exceeding planetary limits.
Source: Author.

the 'Extractive Industries Transparency Initiative', are managed by the industry itself to prevent deeper examination.[8]

Is population the problem?

One aspect of this acceleration is often singled out: population. Isn't this expansion due to rapid global population growth? Aren't there just too many people (and sometimes this thinking goes further and shines a spotlight on poorer countries which tend to have higher population growth rates than the greying Global North)? Let's look at the roles of the poor and the prosperous in overexploitation. Figure 19 depicts in very approximate terms the relative impact of the two sectors.

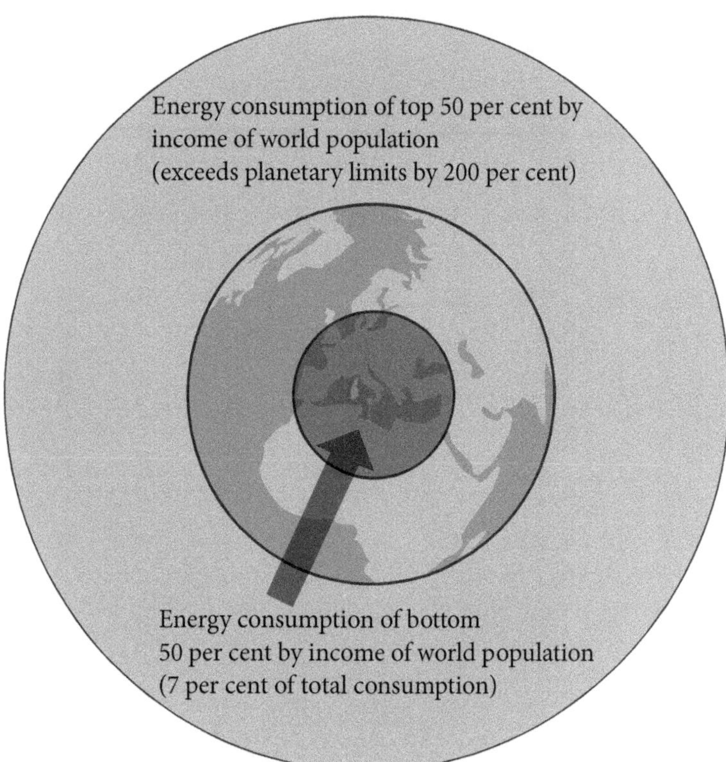

Figure 19 Relative global energy consumption of top 50 per cent and bottom 50 per cent income groups.
Source: Data sources International Energy Agency. 2023 ('The World's Top 1% of Emitters Produce over 1000 Times More CO_2 than the Bottom 1% – Analysis', IEA, accessed 5 January 2024, https://www.iea.org/commentaries/ the-world-s-top-1-of-emitters-produce-over-1000-times-more-co2-than-the-bottom-1.) and International Renewable Energy Agency. 2018 (IRENA, 'Global Energy Transformation: A Roadmap to 2050', 2018, https://www.irena. org/-/media/Files/IRENA/Agency/Publication/2018/Apr/IRENA_Report_G ET_2018.pdf).

As you can see, the overconsumption of energy relative to planetary limits dwarfs the planet, but the equally striking statistic is that the poorest 50 per cent of the population consume only 7 per cent of this total. Similar statistics often focus on the richest 10 per cent, and this group alone are responsible for 48 per cent of total consumption, but I've chosen to include, in very approximate terms, the overall consumption of what I call the 'prosperous' – those who participate in the global consumer cycle, rather than sitting in shanties and slums on the margins of it. The data makes it clear that it's not population that's the problem – it's the prosperous population.

The point of these simple diagrams is to ask where in this cycle is there an entry point at which to move its effects back within planetary limits? Answering that is a major challenge, particularly as not only does overdevelopment of cities need to be managed but doing so needs to take account of improving the lives of the billion or more segregated to the slum/shanty margins – 'leaving no one behind' as the United Nations asserts in the New Urban Agenda and other frameworks.

What is driving the unmanaged consumer cycle? The key role of industry

Innovation and risk-taking are part of the power of private enterprise. It's what has transformed our world in many ways for the better. If the world is to undergo the massive and expensive transition from its current production and consumption system to one which lies within planetary limits it will need all that energy, creativity, innovation and risk-taking to create as yet unimagined solutions. Industry also includes the huge global financial system of banks, investment funds and hedge funds which underpin the industrial machine and finance its actions. But lying behind the industrial complex is a basic drive for existence and self-preservation which we've seen already in oil company strategies to downplay claims of the risk of climate change, and in the huge number of lobbyists at events such as COP28. We can see – in our increasingly polluted, unhealthy and inequitable cities, warming atmosphere and oceans – the costs of this. From the beginning of the Industrial Revolution in English cities such as Manchester, shrouded in smog and acid rain, to the present day the activities of industry have costs resulting from extraction, environmental damage, construction, consumption and pollution. The fundamental problem is that, to return to Stiglitz's phrase, the gains are privatized while the costs are socialized. Economists use the term 'externality' to refer to consequences of economic activity which are not costed in. Since industry's social and environmental impacts on local populations and to the planet are not priced in to profit and loss accounts, industries have no economic reason to reconsider their actions. Throughout our Anthropocene journey we have seen these negative externalities impacting society and the environment. The costs fall on wider society but the companies break no laws. In many other cases however, more malicious, deceptive or even illegal actions are shaped towards self-preservation, from early cases such as the secrecy surrounding 'phossy jaw' in match factories in the nineteenth century to cover-ups of harm resulting from industrial activity such as the infamous case of industrial pollution of the Love canal near Niagara Falls in the United States,[9] the Union Carbide Bhopal disaster, the Deepwater Horizon oil spill or 'ecocide', the Rana Plaza garment factory disaster, the Volkswagen emissions scandal and many others.[10] What 'should' the companies comprising industry do? Is their role simply to innovate and exploit to sustain their operations and generate revenue, or do they have any other responsibilities? What is their place in the capitalist free-market system which appears dominant? What are their limitations?

Industry evades market forces

At the outset of the Anthropocene journey, we encountered individual entrepreneurs such as Arkwright, and later Ford, who acted as risk-takers and innovators to grow their businesses. The setting for their activities was the emerging capitalist free-market structure, reflecting Adam Smith's ideas of the functioning of the 'Invisible Hand' to achieve efficient transactions, optimizing prices. Consumers of these companies' products could shop around in a marketplace for the best price and the best quality, with few other considerations.

As companies have grown in scale and capabilities, developing sophisticated brand identities, products and marketing methods, they have been increasingly able to evade market forces. By the 1920s it was already becoming clear that ideas of perfect competition and profit maximization didn't quite fit what companies were doing. There were fewer and bigger Arkwrights and Fords (and the corporations of Gates, Jobs, Bezos, Zuckerberg and Musk are even bigger). In this new industrial world the products they produced were carefully distinguished from competitors through brand identity and marketing, in the same way as Ford discovered their cars should be sold as prestigious freedom opportunities rather than just machines. They distanced themselves from simple free-market price-based choices through differentiating their products and brands.

The ways businesses were run changed too. Growth and mergers created larger companies. Along with growth in scale came a need for capital beyond that held by company owners, so ownership became diluted by share issues and management of companies increasingly passed from their proprietors to directors and managers.[11] A range of economic theories discuss the effects of corporations becoming managed by executives rather than controlled by their initiating entrepreneurs. One of their main messages is that managers and directors of companies may have other priorities than pure profit. Their own roles, security, status, career progress and income are likely to be important to them, rather than pure profit maximization. The survival and security of the firm become paramount. Herbert Simon coined the term 'satisficing' in 1956, combining the words satisfying and sufficing. It reflected the idea that as long as management achieved a satisfactory profit, that would suffice and they could pursue other goals.[12] As a consequence, the actual goals of the company might be concerned, for example, with stability rather than innovation, to strengthen the security of the managers. Or they might be concerned with market share rather than profit maximization, again to increase security and predictability.

Multinational corporations and market failure

The global explosion of industrial growth after the Second World War added a further scale of complexity to the behaviour of companies as even larger multinational corporations (MNCs) developed. By 2018 there were over eighty thousand MNCs operating worldwide, ranging from oil and gas producers, retailers and car companies to electronics and software producers, finance houses and banks. In 2020 the two thousand largest firms had combined revenues of $48 trillion, equating to sales of $6,050 for each individual on the planet. Mounting evidence shows that MNCs are dedicated to evading market forces and maintaining their own status and position. It also shows trends towards market concentration, reducing the possibility of competition and new entrants. MNCs have also progressively reduced their tax burden, reduced the power of unions and externalized their environmental impacts.[13] The negative consequences of this are described by economists as 'market failures' – ways that failure of markets to *price in* social and environmental impacts results in markets being unable to manage those impacts. For example, Stiglitz and Stern list six market failures relevant to tackling climate change:

- Greenhouse gases – as their production is not priced in to the activities of companies

- Research and development – as companies are not motivated to develop and innovate
- Imperfection in risk/capital markets – limiting availability of capital for green initiatives
- Networks and system change – failure to integrate infrastructure needs for new technologies
- Information – lack of transparency on environmental impacts of products and technologies
- Co-benefits – information on non-economic benefits to global commons and well-being

(adapted from Stiglitz and Stern)[14]

They argue strongly that the ability of corporations to evade market pressures and shape their own agendas, resulting in a prioritizing growth and profit, conflicts with the possibility of them making positive contributions to securing a sustainable future.

Social responsibility?

Several of these market failures reflect ways the market system fails to motivate or release companies to innovate and drive a transition to a more sustainable system. However, the scope – according to recent theories of the firm – for management to pursue their own goals might allow for them choosing to act in a socially responsible way. I encountered a case of this. A UK-based retail company, B&Q, part of Kingfisher MNC, was challenged by a journalist on its ability to account for the origins of its products. They sourced a huge number of products globally. Could they be confident that materials for their products were extracted sustainably? What about the labour practices of their suppliers? The company embarked on a major Corporate Social Responsibility (CSR) programme, auditing its suppliers and supporting them to improve practices. It was a partner in establishing the Forest Stewardship Council (FSC) scheme to certify sustainable timber extraction.[15] The trigger for this was the threat of bad press and damage to the brand, but the driver was individuals within the company with the power to make this change in direction, reflecting their own concerns and passions – at least having worked and travelled with them, that is my strong impression.[16]

Milton Friedman, champion of free markets, disputed this role. He famously said in a 1970 *New York Times* article:

> There is one and only one social responsibility of business – to use its resources and engage in activities designed to increase its profits so long as it stays within the rules of the game, which is to say, engages in open and free competition without deception or fraud.[17]

Friedman's extreme free-market position begs major questions. We have already seen that industry invests huge amounts in lobbying and marketing to manipulate public and governmental attitudes in their own interests, which can fairly be called *deception* or in some cases even *fraud*. We've shown how the power of MNCs, for example, leads them to evade the 'rules of the game'. In fact, in the years since Friedman's rejection of CSR it has been widely embraced by industry and has become big business. Alongside its aims to improve the social role of corporations it can add brand value in the eyes of consumers. At worst it is labelled 'greenwashing' – wrapping green credentials around the brand to reassure the public. The largest global CSR initiative is yet another United Nations project, the 'Global Compact'. Companies can sign up to this and in doing so tick a number of social, ethical and environmental boxes. One study of the effects of membership finds that while many companies benefit financially from membership, not least because it enables them to get on to United Nations vendor lists, it doesn't improve their CSR performance. The root of the problem appears to be that there is no monitoring mechanism, no audit, no sanctions.[18] Yet another industry initiative is 'Environmental, Social and Governance' monitoring – ESG. Companies creating ESG goals claim to be acting more responsibly in the marketplace and the idea has been hyped as a 'win-win' which not only encourages more sustainable businesses but increases their value, though some critics suggest that like the Global Compact the process does more to enhance brand value than sustainability.[19]

Other initiatives have more teeth. The 'Global Reporting Initiative' did push for transparency of companies' CSR performance, though it has now drawn back from publishing data.[20] The idea of the 'B Corporation', managed and monitored by an independent non-profit, is that participating corporations sign up to and are audited on a set of CSR standards. This scheme has been questioned as small companies who were early adopters see large MNCs with questionable commitment to substantive change joining up because of the benefits to their brands.[21]

Despite all these efforts – some effective and some less so – to 'green' industry it is abundantly clear from our historical journey into disaster creation that deep roots of disaster making lie within industry, encompassing extraction, production, marketing and the underlying financial system. Despite the movement towards corporate social responsibility, companies, corporations and MNCs are invested primarily in their own survival. The growing scale and power of MNCs and market concentration enables corporations to protect themselves from innovation and disruption by competitors, stifling innovation.[22] Throughout our Anthropocene journey industry has taken opportunities to privatize gains and socialize risks through:

1. Creation of externalities: Social and environmental impacts which are unmanaged by the market system as they are not priced in and which in many cases extend to deliberate deception and fraud.
2. Market failures: Industry uses a range of mechanisms to evade market mechanisms so that markets fail to maximize benefits for all, pushing their existing building programmes, product lines and energy sources and investing in political lobbying to secure their own survival and profitability.

Government

Faced with unmanaged accumulation of risk through the ever-growing consumption of resources, overdevelopment, environmental and social exploitation, exceeding the planet's capacity, the role of governance is critical in reining in and managing this expanding cycle.

Governance is what governments do. How do they tackle governance of risk creation? In extreme cases their actions are very visible. When the Covid-19 pandemic swept across the world in 2020 they invoked wide-ranging emergency powers. Populations and economies were locked down. Movement was restricted. Spending was diverted onto response, research and healthcare. A disaster gives governments permission to act in sometimes draconian ways, it's part of the 'social contract' between government and citizens.[23]

Getting prepared: Governments and disaster risk reduction

Governments and the international system have also directed their attention at the *causes* of disasters, building institutions and frameworks concerned with disaster risk reduction since at least the beginning of the UN Decade of Natural Disaster Reduction in 1990, but the financial reality of their commitment to disaster reduction and preparedness is very different from their rhetoric. We saw how the UK government appeared to ignore the conclusions of their own pandemic simulations, and actually scaled back their institutional preparedness and supplies in the years leading up to it. Internationally, the report 'A World at Risk' from the 'Global Preparedness Monitoring Board' (convened by the World Bank and World Health Organization) – published three months before the first Covid-19 cases were reported in Wuhan – warned that this lack of preparedness was the case internationally:[24]

> Despite the high cost-benefit ratio of emergency preparedness, governments continue to neglect it. ... Preparedness capacities and systems are global public goods – all countries benefit from every country's investment.

The failure to invest in disaster risk reduction and preparedness is familiar in the world of disaster studies. The idea of the cost-benefit ratio of preparedness compared with response is well recognized. A 2021 UNDRR report suggested that a global €1.6 trillion investment in disaster risk reduction could avoid losses of €6.4 trillion. It contrasted this with an actual spend on disaster preparedness and mitigation from 2005 to 2017 which was less than 4 per cent of the amount spent on disaster *response*. The report suggests that 'Governments do not prioritize disaster risk reduction because they see it as a cost for an event that might never happen.'[25]

Governments allow risk creation to outstrip risk reduction

National governments might be expected to take a key role in risk governance, and yet action to regulate and legislate is slow and ineffectual until risk intensifies into disaster. We've seen that it takes trigger events such as city-wide smogs, financial crashes or large-scale epidemics to stimulate rapid response. From the emergence of the Industrial Revolution factories, legislation to manage working conditions, pollution, emissions, unmanaged development and other risk creation activities has been retrospective, slow and limited. Why?

The war against red tape

We've seen already that 'externalities' – environmental and social costs of industrial activity which are not borne by or priced in by the producers – are not managed by the 'market system'. They lead to 'market failures'. This is where government should come in, managing industrial activity to avoid these externalities. In other words, governments should have a major role in tackling risk creation. The key word concerning government's role in this, according to economists, is 'regulation', but regulation, particularly for a free-market supporter such as Milton Friedman, is simply a burden on free enterprise:

> We have heard much these past few years about using the government to protect the consumer. A far more urgent problem is to protect the consumer from the government.[26]

Friedman's anti-government stance implies that the consumer doesn't need protection. It has fuelled anti-regulation sentiment. While exercising emergency powers – a very dramatic form of regulation – is accepted by many as a role of government, regulation of the ongoing progress of the consumer cycle is often labelled as 'red tape' and distrusted, in line with the views of Friedman and his followers. The UK government, for example, established its 'Red Tape Challenge' in 2011 predating Trump's similar anti-regulation policy, applying a 'one in, two out' approach to creation of any new regulations in which for every new regulation created two existing regulations should be scrapped.[27] The tragic Grenfell Tower residential block fire in London in 2017 led to suspension of this rule[28] to allow any necessary regulations aiming to prevent another such disaster to be created, as it was widely argued that restricting fire safety regulations was a contributory factor in the disaster. In the public enquiry which followed it, counsel to the enquiry challenged the government minister responsible for fire safety at the time:

> The government had an 'ideological presumption' against regulation.[29]

The see-saw from rejecting regulation through to 'red tape challenges' and to reintroducing it after the tower block disaster is one which occurs in many situations, for example, in the introduction of regulations in the United States to separate investment and commercial banking during the Great Depression, their repeal in the years leading up to the Great Recession, introduction of new financial industry regulations (the Dodd-Frank Act) after that event and relaxation of those regulations in 2018.

The idea of externalities

The colloquial term 'red tape' is misleading,[30] implying government rules that shackle business and add costs. The term suggests that regulation is always burdensome and unwanted. In fact, both industry and the public often welcome regulation. For example, industry welcomes regulations that protect their intellectual property rights, restrict market entry, refine corporate law and manage the financial environment so that they can do business safely, predictably and profitably.[31] However, it resists regulation of negative externalities such as environmental and social impacts of industrial extraction and manufacture. We've seen that these costs are not 'priced in' and some argue that there should be mechanisms to do so, an idea first developed by economist Arthur Pigou over a century ago.[32] Where industry can exploit environmental and social resources and cause damage and degradation but does not bear the cost then those costs are *external* to their operations. But they represent a cost to someone – to those whose lives or environment is affected. One estimate is that the global negative social and economic externalities of industry actually cost the global economy US$4.7 trillion per year,[33] equating to 13 per cent of global economic output. The scale of this economic 'free ride' is such that if some companies had to pay a fair price for this exploitation they'd go bust.

Many, taking their cue from Pigou's work, argue that mechanisms such as taxes could make them *internal* to industrial activity. In other words the full cost of business operations should be borne by them, leading them to think again about the negative environmental and social impacts of their activities. 'Carbon taxes' have been widely discussed as a means of making industry take responsibility for its carbon emissions. More generally such taxes are referred to as 'Pigouvian Taxes'.

The trick is that industry can exert huge influence and even enlist public support where regulation is not in industry's interest. The Anthropocene case studies demonstrated industry resisting regulation of externalities to their activities – for example, pollution, environmental damage and labour exploitation. I referred earlier to the governmental tendency to 'see-saw' in the application of regulation. Given the contrasting and ideologically tinged views of its role it is easy to see why. The forces that resist it tend to hold sway in times of stability when danger seems remote, and those that welcome it dominate when disaster strikes: 'Depression and war give us high taxes and regulation.'[34] Carbon taxes, though widely discussed, have not generally been imposed. Not only does industry resist this added cost to their operations, but the public may also resent higher charges for energy and transport, for example. Governments in turn respond to public attitudes as well as industry pressure and lobbying. The true costs of the lifestyles the prosperous enjoy remain hidden, at least in the short term.

What about global governance?

As well as national governance we've examined the role of the international system, particularly represented by UN bodies, as many aspects of risk creation and magnification don't respect national borders but are international in scale. What about governance at the international level? Can't these global impacts be tackled globally? We've already recognized that governance at this scale is limited. Why is this? Behind the highly visible actions of UN agencies there's a lot of discussion about this challenge. 'Think pieces' generated within the UN grapple with it. A group based in UNDP produced a substantial book: *Global Public Goods: international cooperation in the 21st Century* which concluded that there were three major challenges to global governance:[35]

1. A *jurisdictional* gap, in which global organizations don't have the ability of national governments to enforce policies.
2. A *participation* gap in which the growing number of non-governmental global networks and organizations are not properly involved in global governance.
3. An *incentive* gap, in which national governments are not sufficiently motivated to cooperate in global policies.

We have seen already a jurisdictional gap in the example of negotiating the SFDRR disasters treaty, where national concerns trump global policies. We've seen, further, that implementation of frameworks, for example, goals of the climate treaty, is inhibited by the lack of global jurisdiction, trumped again by both governmental and commercial concerns. The participation gap refers partly to failure to engage actors such as civil society organizations, but the writers also acknowledge that heavyweights such as the IMF, World Bank, G20, G8 and World Economic Forum often have more jurisdictional power than the UN's framework-creating agencies. Money talks!

Finally, while the need for incentives for national and commercial participation is clear, what is unclear is what incentives would be sufficiently powerful to secure cooperation.

The UN produced a further 'think piece' while developing the three key 2015 frameworks. 'Global governance and governance of the global commons'[36] imagined a future in which the concerns of the frameworks were completely integrated and were backed up by stronger global governance. The report did not, however, address the jurisdictional, participation or incentive challenges of the UNDP analysis mentioned above and subsequent events have shown that these gaps remain, leading to a lack of integration of the frameworks and continuing lack of effective global governance of them.

The underlying challenge to the effectiveness of the UN can be seen as ideological, reflecting a battle between 'Realists' and 'Liberals'.[37] Realists argue that nation states act out of self-interest and, lacking a global scale of government, they operate 'anarchically' in the international arena. Liberals, or idealists, argue that actions are also motivated by values and this gives scope for international cooperation based on commonly shared values. While UN thinking embraces liberalism optimistically, often global governance outcomes owe more to a realist perspective.

As we've charted the role of UN institutions in attempting to manage our planetary global commons sustainably, its limitations in doing so have been very visible. The UN system has had many successes, for example, in establishing internationally accepted norms, encouraging international cooperation, supporting emerging states and strengthening humanitarian response.[38] But this assessment of their scope for global governance concludes that they are unable to do much in the face of national and commercial interests.

The roots of disaster making

The simple cycle we've used to relate industry, government and the public in a process of unmanaged growth represents a *system*. While individuals and organizations are sometimes clearly responsible for disaster making there are much wider systemic forces at play. All three elements of the consumer cycle play a part in disaster making (though we are yet to address the role of the public). What are the roots of this system? To answer this we need look back even further than the starting point of our Anthropocene journey.

We've mentioned briefly that the social and philosophical shifts of the Enlightenment were important pre-conditions for the emergence of free-market capitalism. In this new era the ideas of private ownership, of the management of commerce through an 'invisible hand' rather than authoritarian control and the possibility of accumulation of private property and wealth led to the trajectory of growth, innovation and industrialization which spans the Anthropocene period. Advertising and consumerism are built on this underlying ideology of individual freedom and free markets.

There are grounds for regarding the idea of 'free markets' as an ideology, by which I mean a set of ideas rooted in a belief system which lies beyond rational understanding. I describe the idea of 'free markets' in these terms as it conflicts with clearly visible evidence. It's generally recognized that sophisticated markets *aren't* free because corporations make sure they're not. We've seen how the scale, managerial structure, branding, marketing and lobbying activities of corporations enable them to evade market forces. Economist Kate Raworth said, 'Whenever I hear someone praising the "free market" I beg them to take me there, because I've never seen it at work in any country I've visited.'[39] Corporations themselves welcome interventions limiting the freedom of markets when it benefits them, and more generally all sectors of society welcome government interventions when they are in their interest. Nevertheless they champion freedom. Ironically the growth of the free-market economy in the United States was fuelled by government intervention in the form of protectionism in the early years of the twentieth century. Nevertheless surveys, as well as voting patterns, continue to show that the public welcome and support 'free markets'[40] and tend to vote down regulation ('red tape'). This behaviour is ideological, rooted in social and philosophical shifts dating back over three centuries. A fascinating confession comes from Bill Gates, who built one of the world's biggest corporations, Microsoft, writing about why he now sees government intervention as vital in changing track to a sustainable future:

> It might seem ironic that I'm now calling for more government intervention. When I was building Microsoft, I kept my distance from policy makers in Washington, D.C. and around the world, thinking that they would only keep us from doing our best work. In part, the U.S. government's antitrust suit against Microsoft in the late 1990s made me realise that we should've been engaging with policy makers all along. I also know that when it comes to massive undertakings – whether it's building a national highway system, vaccinating the world's children or decarbonizing the global economy – we need the government to play a huge role in creating the right incentives and making sure the overall system will work for everyone.[41]

Even billionaire Gates recognizes that the ideology of freedom and free markets increasingly flies in the face of reality.

Common ground

The ever-expanding cycle of consumption embracing the world is unsustainable, pushing beyond planetary boundaries. I've made the case that the dominant ideology of free markets and its deeper principle of individual freedom lies behind this. An economist friend I discussed this with said, 'what's the alternative, communism?!' Beyond that binary I think an alternative which *is* worth considering is the idea of the 'commons'.

This is an idea with history. In many countries areas of land and water are, or were, 'commons'. In the UK much land was held in common. Though often managed by landlords and gentry there were 'commoners' rights' to utilize it. People voluntarily restricted their individual freedom to exploit the commons for the benefit of all, and ultimately of themselves.

In the UK, moves to enclose and privatize land accelerated at the time of the onset of the Industrial Revolution. Major 'Inclosure Acts' (*sic*) were passed in 1773 and 1845, moving land into private ownership, often coalesced into larger holdings.[42] The privatization of the commons was underpinned by the liberal philosophy of thinkers such as Locke and Grotius[43] and as with Adam Smith's 'Wealth of Nations' relatively simplistic understandings of their thinking created part of the intellectual and moral justification for this transition. The idea of the commons was lost in the rush for private ownership and profit.

One result of concern about continuing global risk creation is rediscovery of the idea of the commons, taking it to a global level. 'Common Pool Resources' – such as the atmosphere and the oceans – aren't privately owned but are resources held in common for all.[44] The protection of global commons such as these is clearly an aspect of the international frameworks we have been considering.

These commons are suffering. We noted an estimate of the global social and economic cost of exploitation of the commons by industry at US\$7.3 trillion per year.[45] Externalities have been mentioned several times in our journey and this statistic spotlights them at a massive scale. Since our cities and our planet are our home don't we all have a stake in them? Should industry be allowed this seven-trillion-dollar free ride? Don't we have some right to protect our commons?

A strong argument *against* this sense of common ownership is put forward in Garrett Hardin's *Tragedy of the Commons*. He claims it is inevitable that any resource held in common will be increasingly exploited as those making use of it expand their numbers and their operations.[46] He gives the example of the planet itself, with its population growing, he argues, unsustainably. His conclusion is that only if all resources are privatized will they be managed. As with any theory his depends on his starting assumptions about human behaviour. He, like many economists, takes self-interest as a primary motivating factor. Economist Elinor Ostrom presented a counterargument, reflecting her distrust of abstract theory. In her book *Governing the Commons* she turned to case studies rather than theory and searched for different situations, historical and current, where common pool resources were exploited.[47] Her studies at local and regional level showed people had the capacity to self-manage for the benefit of all, because not doing so would result in the kind of tragedy Hardin predicted. Ostrom went on to study, discuss and promote the idea of the commons more widely. She recognized the distinction between managing commons at smaller-scale levels and the challenges at national and global levels (where trust, participation and collaboration are much more difficult to achieve) and reflecting on this challenge she also argued for 'polycentrism' – different levels of governance with different responsibilities.

As city land is buried under concrete and tarmac the environment and the planet are easily forgotten, or at best seen to provide 'ecosystem services', just another resource to feed the system. Thinking instead of our environment as a commons to be respected, cared for and not over-exploited seems a better way forward. Some even speak of the 'rights of the planet' and the 'rights of nature'. There's a link here to basic human rights, as many of these depend on the planet itself, so that if its rights are not protected and it is progressively degraded then human rights, in turn, will be undermined.[48]

The city is a specific focus for 'commons' thinking. Though much of the city is privately owned, it is home to everyone who lives there. The idea of the 'urban commons' provides a strong basis for negotiating development of hospitable, prosperous and equitable cities.[49] The idea can be visualized at a number of scales. At the smallest scale it can be seen in the local disaster reduction stories recounted in Chapter 1 ('Local Heroes'). The account of the community association in East Delhi shows poor residents in this informal area exerting influence over their environment and over the municipal authorities, claiming rights to their urban commons. Ostrom's work and that on the urban commons suggests that the local scale is an important starting point. Phenomena such as 'street level innovation'[50] also highlight the agency of local actors in driving change. At a broader scale activities to take control of wasteland, brownfield sites and other unused areas enable local organizations to create useable green and activity space. The development of relationships between local CSOs and municipal government, as seen in the East Delhi example, is important in extending influence over the urban commons as government – if it wishes to and has the capacity – can manage not only common pool but private resources through regulation. At the city level and even more widely, scenarios for sustainable futures depicted by UN frameworks require development of societal responsibility and collaboration to effectively manage the exploitation of planetary resources, and that, in its most basic sense, is the idea of the commons.

Ostrom's case studies of governing the commons showed that 'rule-breakers' were sanctioned for everyone's benefit. She acknowledges that these principles of governing the commons work at scales where people are conscious of their impacts on their local environment and are able to collaborate to protect it. At a larger scale this awareness and collaboration is much less likely. Nevertheless we have seen that while in the earlier phases of development negative externalities could be absorbed and ignored the impacts of the 'great acceleration' force us to consider how to protect and sustain our cities and our planet. So who might act, and how, to protect the urban and global commons?

Who will protect the commons?

The graphic representing this expanding, unsustainable exploitation of the global commons looks a little like the window of a washing machine during the spin cycle (Figure 18). But unlike that machine there is no simple and obvious, and certainly no instant 'off' button to the risk-creating consumer cycle.

It seems clear to me that there is only one, hugely problematic entry point at which this cycle can be broken. The underlying driving forces of industry disable it from dramatically changing tack of its own volition. The competing pressures on government from the public and industry, within short-term electoral cycles reinforcing short-term thinking, disable it from substantive action. The only possible entry point is the *public*. I say this is hugely problematic because the track record for publicly driven change isn't always good. We have charted essentially passive public behaviour during our Anthropocene history, and shown how people express preferences which discount the consequences of their choices. And yet the public are in a position to make choices about the products they purchase, about how they travel, about how they live. Those choices, if made differently, affect industry. The public can also express their preferences and vote in line with them. Within democracies this affects in part which party is elected and more significantly what policies parties put forward to seek election. The public can also, if they choose to, organize through campaigns, through formal civil society organizations, through unions and increasingly through social media. Representatives and leaders within industry and government are also members of the public and may choose to act for the common good (I gave the example of individuals within the B&Q business championing corporate social responsibility). The public are those who inhabit the urban and global commons, and we've already seen examples of the public taking action at very local level to protect the commons. How might they take action at some of the other scales necessary if that 'stop' button is to be successfully pressed? How might they become 'changemakers'?

Chapter 12

THE PUBLIC AS CHANGEMAKERS

If people find that their lives are threatened, and that they are being exposed to risk creation which will affect both themselves and their children, won't they speak out? Won't they take action? How do the public approach threats from risk creation?

Where we've noted public attitudes in our Anthropocene journey, we've found examples of acceptance and passivity in relation to risk creation and its effects. Amid the belching chimneys of the Industrial Revolution a resident welcomed the sight as assuring them of a warm hearth – of prosperity. In today's cities, similarly, prosperous inhabitants seem resigned to the pollution and health effects of the smog and fumes clouding and clogging their cities, for example, Bangkok residents preferring using their cars to getting on the Skytrain. The prosperous have power and choices but faced with these threats they seem passive, or at least trade-off the costs for the benefits of … what? Autonomy and freedom? We've found, for example, that from early Ford advertising onwards the car has been associated as much with aspirational values of freedom as with practicality. 'Social capital' reflects a presence or absence of community cohesion. In many societies this quality is diminishing.[1] For the poor, lack of resources, options and power constrain their ability to act, and sometimes inevitably (as in the cry of the Honduran campesino that 'I am destroying the land') lead them to accepting an accumulation of risk. For the prosperous it seems that there is a willingness to accept the externalities associated with industry and consumption. A preference for personal freedom is *dissonant* with knowledge of the consequences to their own lives and future generations. In this sense, as willing consumers of the products, services, transport and energy, the public, co-opted into the system as a vital part of it, are partners in the unsustainable consumer cycle.

Cognitive dissonance

How do the public hold two contradictory facts in their heads as they seem to in participating in an unsustainable consumer cycle? One explanation is the idea of 'cognitive dissonance'. An example of this is a study which interviewed current and former oil company executives. As employees in businesses clearly implicated in greenhouse gas production, how did they justify their involvement in these companies? The study found a clear distinction between interviewees who had left jobs in the oil industry, 'leavers', and those still working there. Leavers were clear that there had to be change, one, for example, said, 'The oil industry will have to decarbonise by 2050 whether they like it or not.'

The language of executives still at the companies was much more ambiguous. They talked of 'green business' and becoming 'integrated energy companies'. Despite a general recognition of the need for change they defended their role, one, for example, arguing, 'No one has the right to tell someone they can't have energy' (implicitly making 'energy' fossil energy), and in some cases deflecting blame: 'energy companies get the torchlight pointed at them. But how willing is Joe Bloggs down the street to give up petrol cars, or have a heat pump?'[2] The researchers allied the views of their interviewees to the phenomenon of 'cognitive dissonance'. This is the idea that when two conflicting ideas have to be held in tension people rationalize their ideas and beliefs to allow them to pursue their preferred choice. Those working at the companies, wanting to retain their roles, used narratives which rationalized or deflected the consciousness of the global impact of their industry. Those who had left were no longer in this dissonant position and were able to recognize the necessary changes.

Cognitive dissonance among the general public may be one explanation of the contrast between information on consequences to themselves and future generations of unsustainable development and of their individual lifestyles, voting choices and even their investment choices despite this information. For the interviewed oil company executives, the tension was between role and career on the one hand and the consequences of their companies' activities on the other. For the public the tension appears to be created in part because of their expectation for freedom. The nature of the consumer system is to provide the prosperous with huge opportunities, options and freedom to choose between them in the same way as the free-market philosophy lying behind the industrial system confers freedom to innovate, create and exploit with limited responsibility for the negative environmental and social impacts – externalities – of its actions.

The public – Possibilities and paradox

What can we expect of the public? As I write, tractors are blockading city streets in Germany. Farmers are protesting at the reduction of subsidies on diesel fuel.[3] The government has already scaled down the changes but the tractors continue to disrupt traffic. The farmers are angry. Ironically dramatic storms and floods in Germany in the previous summer,[4] impacting farmers as much as anybody else, may be early warning of increasing extreme weather events driven by global warming, a result in large part of fossil fuel consumption … like diesel. I'm told by a colleague in Germany that campaigning by climate activists is treated much less sympathetically by the media (we will consider further how media affect public perceptions). The news makes me despondent about the possibility of society working together to protect the commons and keep our world sustainable for us all and for our descendants.

In our own town, just a few weeks ago, tractors were out on the streets too. A procession of fifty or more of them during the Christmas celebrations, decked with decorations and lights, processing through the main streets. This wasn't a protest; it was a fund-raising drive. They were inviting everyone who applauded their procession to donate to charity.[5]

What can we expect of the public? Farmers and rich urban commuters are just like the rest of us. Often preoccupied with their own problems, but also playing their part in caring about the wider society. But is it possible for society at the scale of nations, let alone the whole planet, to act together to protect its common property, the Global Commons, its one and only home (until of course Elon Musk jets off to the stars)?

Everyone loves a good conspiracy theory. I'm not about to offer you one, but I *am* going to suggest that not everything's as it seems with the public. There's a lot of material on public attitudes to climate change so in this section we'll examine attitudes to climate change to assess the scope for public concern and action on this and on more generally changing course towards a sustainable, prosperous and equitable future.

A friend of mine, a climate sceptic, said 'once people realise all this stuff's going to cost something they'll lose interest rapidly'. Politicians make a feature of siding with the climate sceptics, believing (presumably) it will win them votes. Brazil's President Bolsonaro encouraged the continuing devastation of the Amazon. President Trump withdrew the United States from the Paris climate agreement. In the UK Prime Minister Sunak backtracked on emissions reduction targets. The thing is, there's strong evidence they're wrong about the public. Wide-ranging surveys show not only that a majority of people are concerned about climate change but also that (despite my friend's views) in some cases they are prepared to accept costs and changes in lifestyle to do something about it.

So let's scan a few of the findings from attempting to establish what the public *actually* think.

A UK public survey of 5,665 respondents (2021) found overall public support for policies addressing climate change, but also found that support dropped substantially when people were told that such policies would affect them personally, for example, by increasing prices and taxes.[6] It noted a divergence of views between those right-wing and left-wing voters, with the latter more supportive of policies addressing climate change. What strikes me about this survey is that while it highlights the costs of the various actions, it doesn't set against that the costs of *inaction*. One of the survey organization's directors echoes this point:

> This support provides further evidence that the UK public want urgent action on climate change, but may not yet be fully aware of the implications for individuals of doing so. There is a need to raise awareness of the costs of action but also of inaction.[7] (author's emphasis)

An international survey reached rather different conclusions. It consulted over sixteen thousand respondents in sixteen advanced countries (2021).[8] Most strikingly the study found that over the whole sample there was a strong majority (80 per cent) willing to make changes to how they live and work to help reduce the effects of climate change. In this case the phrasing, directly linking *costs* to the *benefits* of reducing effects of climate change, may have led to the more positive response to bearing personal cost. In all countries studied those on the ideological left were more willing to adjust their lifestyles in response to climate change than those on the right. The divergence was particularly striking in the United States, the only country in the survey where right-wing respondents expressing concern were in a minority. To the extent that right-wing ideologies emphasize personal freedom this supports the idea that resistance to climate change action is linked to holding onto personal freedom. There was a marked difference also between the attitudes of younger and older people, young people showing more concern about the problem. The same survey found that intense concern about personal effects of climate change had increased sharply in several of the major economies surveyed between 2015 and 2021, falling only in Japan.

The main finding of these surveys – that a majority of respondents are concerned about climate change – is replicated in surveys in the United States (63% of 1,278 respondents, 2018)[9] and in Europe (77% of 26,358 respondents, 2023).[10] A very-large-scale 'light touch' survey conducted globally via phone apps by the UN found that 64 per cent of 1.2 million people surveyed in fifty countries felt that climate change was a global emergency (2021).[11]

A larger survey as part of the Gallup World Poll (2021/2) asked whether the public would bear a specific cost of climate change mitigation. It consulted thirty thousand individuals in 125 countries and found that 69 per cent of respondents were prepared to contribute 1 per cent of their personal income to support mitigation of climate change.[12] This and the other international survey quoted above suggest that public concern extends to accepting at least some cost in transition to a more sustainable future. But there is a paradox here. Despite the data, the continuing impression is that the public remains passive.

The paradoxical public

It's clear from all these surveys that a majority of people are concerned, but less clear whether they are prepared to take action. A UK study looked more closely at a range of attitudes to climate change, based on 31,498 responses (2018–20).[13] As well as identifying groups who were concerned and those who were sceptical it found that 40 per cent of respondents were what they termed *paradoxical* in that they tend to be worried about climate change but feel powerless to act on it. They report similar findings in comparable surveys. We've already encountered a similar idea: cognitive dissonance, illustrated by the contrast between attitudes of current oil industry executives and their colleagues who have left the industry. Both paradoxical thinking and cognitive dissonance suggest that people tend to hold two contradictory attitudes in tension and are frozen into inaction.

Various ideas have been put forward to account for this substantial proportion of the population who are disengaged from action on climate change even though they recognize it is a problem. They include lack of relevant information on options for action and the sense that concern is not shared by a majority.

Knowledge of positive options influences attitudes

One study found that people were more likely to be concerned about climate change if they were made aware of constructive, positive actions taken personally, publicly or politically which would mitigate its effects. The researchers found that awareness of options for positive action actually drove concern about climate change, rather than the other way round.[14] Another commentator pointed out that such information, particularly on personal involvement, has been markedly lacking compared with public information efforts in other crises such as wars and the Covid-19 pandemic.[15] An experiment on information and attitudes to climate change mapped the attitudes of participants at the outset of a 'world simulation' in which they were provided with information on upcoming climate impacts and had to make decisions about what action to take.[16] The study of 2,080 participants (2015–20) found that after the simulation the participants had shifted their attitudes towards supporting action for change, and that this shift was most marked among those who were initially sceptics. It seemed that getting closer to the facts affected peoples' views.

Knowing what others think influences attitudes

Attitudes have a social dimension, and another study identified the phenomenon of 'Pluralistic Ignorance' (academics love to come up with opaque terms).[17] This translates as 'You may think no one thinks as you do, but actually they do'. They found that people who were concerned about climate change believed that only a minority shared their views. In fact, as I've shown, a wide range of surveys show a *majority* share their views. The study went on to suggest that people feeling they are in a minority tend to keep their heads down, keep their views to themselves, feeling marginalized. Politicians in turn 'read' their public as being unconcerned.

Why is this? Perhaps it's something to do with those tractors. I felt demoralized by that protest, thinking 'Is this what everyone feels?' You could say I was suffering from pluralistic ignorance. The researchers who put forward that idea suggested that the disproportionate space given to extremist views was one reason for this misconception (it seems likely from our survey data that a majority of Germans wouldn't share the views of the tractor drivers), and another study argues that the way that the public are treated in reporting about politics and other matters tends to portray them as passive and unconcerned. The typical selection of a few 'vox pops' (media industry jargon for *vox populi* – voice of the people) doesn't attempt to engage them as it would a politician on a current affairs programme, but just collates a few 'sound bites' failing to capture any depth to public thinking and attitudes.[18] The lack of in-depth public voice in the media reinforces the sense created by 'pluralistic ignorance' that there isn't a majority of people out there who are concerned and who want to see action, even though surveys show that there is.

Passive consumers or active citizens?

Jon Alexander's book *Citizens* suggests that part of the problem is the use of language. People who are regarded as consumers perceive themselves differently if they are described, instead, as *citizens*.[19] I think this point is really important. Founders of community-based change processes such as Paulo Freire[20] and Robert Chambers[21] took the starting assumption that people could become changemakers rather than having change done to them. They could become active citizens rather than passive consumers (or non-consumers). I am much more interested in this potential – for example, illustrated in the stories of 'Local Heroes' (Chapter 1) than in behavioural approaches such as 'nudge' which take the methods of marketing to shift peoples' attitudes and actions in various ways.

If the public are concerned and potentially able to act as active citizens rather than passive consumers, what changes would they champion? How would they become changemakers? Is it about championing the *means* of making changes which make our world more sustainable, or the *goals* of doing so (possible futures)?

Routes to the future

There's a spectrum of ideas out there about the *means* – political, social, economic and technical changes – that might create pathways to a sustainable future. They range from 'techno-fixes' which preserve our current economic and social models but claim to fix them so that they are sustainable, to revolutionary changes in the economy and the society such as 'degrowth'.

Business as usual	Techno-fix	Green new deal	Doughnut economics	Prosperity without growth	De-growth
Position adopted by industries and governments whose policies in practice maintain degradation of the Global Commons	Allowing business as usual by using existing and novel technologies to reduce and mop up emissions	Large-scale investment in shifting economic development towards a sustainable and equitable green model	Redefining governmental economic goals away from a drive for GDP to operating within a sustainable 'doughnut'	Rejecting the growth paradigm and reconstructing an economy shaped for a broader socially focused sense of prosperity	Transformational shift of all dimensions of economics and society towards a lower consumption model

Figure 20 Spectrum of options for transition to a sustainable future.
Source: Author.

The spectrum of options illustrated in Figure 20, from status quo to radical transformation, reflect a range of industrial, political, economic, technical and social actions. All except 'business as usual' and the 'techno-fix' option are concerned with sustainable and equitable futures. The diagram and brief summaries below depict the social, political, economic and technical spectrum of thinking. This account is very brief, as I will go on to show that I think it's better to start with clear visions for sustainable futures, and then to work back to the means for achieving them.

Business as usual

Negotiations such as those surrounding the UN frameworks we have surveyed reveal agents resistant to change, including, for example, governments, in relation to aspects of the Sendai Framework for Disaster Risk Reduction, and the oil industry in relation to COP28 objectives. It seems that for these entities, perhaps reflecting cognitive dissonance, short-term gains outweigh long-term costs.

Techno-fix

One aspect of resistance to pursuing the targets proposed at the COP28 conference is through promotion of technologies which mop up emissions after they are produced ('Carbon Capture and Storage'). The strategy of dealing with waste, emissions and pollution after their production by dispelling, dissipating or storing it is one that we've seen since the time of chimney-filled cities in the Industrial Revolution. Some approaches to technical fixes are more thoroughgoing, such as Bill Gates's approach to development of technologies to assist decarbonization, allied to personal investment in innovative industries to develop such solutions.[22] Overall, his approach is to use technology to allow us to continue on our current track, but sustainably.

Green new deal

Echoing the idea of Roosevelt's economic stimulus package designed to dig America out of the Great Depression, the Green New Deal (GND) combines decarbonization of the economy with a stimulus package to drive that and also the improvement of society through job creation and reduction of inequality. There are a number of variants of this idea, argued for instance by Ann Pettifor[23] and adopted by US politician Alexandria Ocasio-Cortez, and it has also been used as a way of packaging a wide range of policies in the same way as we've seen that 'Corporate Social Responsibility' (CSR) can add green credentials to businesses and products. This leads to the criticism of both approaches, suggesting that they can add up to little more than 'greenwashing'.

Doughnut economics

The 'doughnut' is a more thoroughgoing approach to redefining economics for a sustainable future based on a diagram and conceptual framework developed by Kate Raworth, setting national and global economic trajectories which sit between the outer wall of the doughnut, beyond which we are gradually destroying our global commons, and the inner wall within which people do not equitably enjoy prosperity.[24] The underlying message is that the economics of a century ago is not fit for purpose and that alternatives such as the doughnut form the basis for redesigning how the economy and society work.

Prosperity without growth

'Doughnut economics' author Kate Raworth describes herself as 'growth agnostic', whereas Tim Jackson, the lead author of the UK Sustainable Development Commission report 'Prosperity without Growth', pushes for the necessity to move away from economic growth, redefining societal goals and focusing on other pathways to prosperity which are inherently sustainable.[25] This reflects much other thinking which draws the connection between unsustainable growth and models of economic management which depend on such growth.

Degrowth

A more radically transformational approach developed over several decades and championed recently by Jason Hickell and colleagues,[26] degrowth calls into question many of the existing dimensions of society and its underlying economic paradigms, proposing a socio-economic structure not driven by material consumption. It rejects the capitalist model and, rooted in ecological economics, proposes a societal shift towards communal values. This is a more overtly political approach than others mentioned (though as I have said before, everything is political).

The costs of change

It's understandable, given growing concern about the unsustainable consumer cycle, that the focus should be on the *means* of change, on technical and economic models which address unmanaged and unsustainable growth in consumption, exploitation, emissions, pollution, environmental and climatic degradation. As one moves towards the right-hand end of the spectrum the ideological battleground around these ideas intensifies as the implications for individual lifestyles as well as on political structures and on industry become substantial. Any change comes at a cost so there is great appeal, towards the left-hand end, in approaches which preserve and fix the status quo. However, the accusation of 'greenwashing', levelled at CSR and GND approaches for example, raises the question of whether such approaches can really address the current overconsumption and degradation of our urban and global commons. Bill Gates raises a further problem when he identifies the roots of his own concern, which lie in witnessing the effects of lack of access to energy in the developing world while working on anti-HIV and anti-malaria programmes. If approaches such as his or GND are going to create an equitable transition they have to deal with the *increased* demands from millions in cities and rural regions who currently don't have equitable access to energy and resources.

The economic, political and social implications of transition are clearly huge, whichever path to change is adopted. Not surprisingly, as a result there is industrial and governmental resistance to change, as well as public resistance to the costs of transition: both economic and 'experiential' (changes in lifestyle, for example), leading to strong preference for the status quo. While there's a lot of work on the economics of managing 'abatement transition costs' enabling businesses to cope with these and protect their profitability,[27] the public are more concerned with what the costs of change are to *them*, and therefore what *benefits* would outweigh those costs.

Figure 21 represents a scenario of transition towards sustainability in our urban and global commons. It suggests that the substantial costs of transition lead ultimately to greater benefits than costs. For example, there are *economic* transition costs of massively improving public transport so that the public willingly relinquish private car use, and also *experiential* costs of personal change towards a different lifestyle in line with stable, sustainable prosperous and equitable environments. For example, to what extent are people prepared to embrace plant-based

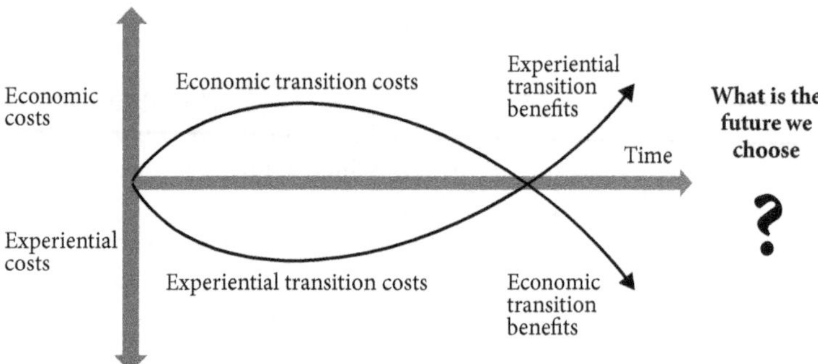

Figure 21 Economic and experiential transition cost curves.
Source: Author.

diets, knowing that meat production massively increases the climate impact of agriculture? It's been recognized that transition costs can fall unequally too, potentially impacting further fragile low-income groups, leading to calls for a 'just transition'.[28]

The diagram suggests that the *economic* curve of transition costs leads ultimately towards economic benefits. It also suggests that the *experiential* costs, emotional and psychological costs of change, lead ultimately towards a more prosperous and healthy future. Analysis such as the Swiss Re report referred to earlier shows that the costs of *inaction* will accumulate. They estimate that by 2050 a 'business as usual' approach to global warming will result in a 10% downturn in global GDP. But there's no getting away from those humps, the costs of change. The scale of a transition necessary to address risk creation, drive it down and establish a society living within a sustainable global commons is massive. I mentioned that I think defining goals is a more effective starting point than focusing on means. As the diagram suggests through the big bold question mark, the underlying question is *What is the future we choose?* Remembering one study which found that presenting positive options for the future was more influential with the public than presenting the problems, it seems to me that this question is fundamental. 'What are the *goals* that would justify those transition costs? What future scenarios would be sufficiently attractive to motivate action? Two of the architects of the Paris agreement on climate change asked this question, titling their book on possible futures, *The Future We Choose*.[29]

From costs to benefits – Vision for the future?

Is the future we choose the same as the present, but worse (as the Swiss Re report calculates)? I'm reminded of 'boiling frog syndrome' – the claim that a frog swimming round in a pan of water which is being heated up will continue to adapt to the change in temperature until eventually it boils. As I've travelled to global cities and mega-cities and seen how their environments are progressively degraded, for prosperous as well as poor citizens, I wonder how long people will put up, like the frog did, with their declining and unhealthy environment. I'm led to think even more strongly that the goals which should be the basis for action are *positive* alternatives to this passive acceptance of worsening and unhealthy environments – *futures we would choose.*

The future we choose

What would positive futures look like? The book *The Future We Choose* is subtitled *The Stubborn Optimist's Guide to the Climate Crisis*. Its authors are Christiana Figueres, executive secretary of UNFCCC in the period leading up to the Paris Agreement, and Tom Rivett-Carnac, who she hired as chief political strategist.[30] As the book's subtitle suggests, they have a hopeful perspective. They defend this as a more constructive approach than 'paradoxical' attitudes which combine concern with inaction. In their view the key to progress is through defining the goals we want and moving towards those goals, gradually tackling the challenges involved in achieving them. Those challenges may seem complex, even unsurmountable, but in their view the starting point is to decide, optimistically, to work towards the future we choose, and as with many social movements the seemingly unsurmountable hurdles can be gradually overcome. Paulo Freire titled a book *We Make the Road by Walking.*[31]

All of which begs a question. What is the future we choose? It's not about numbers calculated by the IPCC, measures of trends in pollution, emissions, growing climate emergency, environmental degradation, the overexploitation of our Global Commons, the persistent creation of risk. It's about what we choose instead, what future we want. It's about more of, not less than.

Although Figueres and Rivett-Carnac are deeply concerned about climate change, what strikes me as they set out to depict futures is that they are concerned not just with climate but with people, prosperity, equity, sustainability, risk reduction and peace. Their aim is to drive down risk and redefine how our world works sustainably, for the benefit of all. What would that look like?

Future cities

They focus on cities in exploring futures we might choose, echoing work done by the UN Habitat 'New Urban Agenda' for cities,[32] and other thinking over the last thirty years by urban planners, architects and others on 'compact cities'.[33] All this work aims to define a positive future that we might choose for cities, one that reverses their overdevelopment and pollution and creates prosperous, sustainable and inclusive environments: *a future we would choose*. The following scenario draws on this work to sketch out a vision for ways cities could be refreshed and transformed as part of a sustainable global commons. This will demand huge investment and in many cases the redesign and retrofitting of existing building, land and infrastructure. However, urban land expansion of between 70 and 171 per cent is estimated to occur between 2015 and 2050[34] so there is huge potential for managing this huge new development, yet to be undertaken, as part of the transformation of cities.

> It's early morning, a fresh clear day in the early spring of 2050. Coffee in hand, you go to your apartment window. Looking down from the third floor a broad strip of greenery and pathways stretch away in both directions, with a narrow service road on the far side. You open the window and hear the conversation of parents and children heading to the local school. They chat together as children dodge in and out of the hedges and shrubs. There's an all-pervasive sound of birdsong. There are trees everywhere, and you remember learning at school what a significant part tree planting played in the global mission to turn the world around. You spend a few moments breathing in the cool, crisp air and remember that your elderly neighbour explained this green strip was once a fume-filled urban highway, until the transition to good quality public transport, along with localized work, shops, services and facilities shrank the need for cars and trucks. Like all compact cities and mega-cities yours has its own style and culture but shares this peculiar combination of high density and open space.

The decarbonization of the city and the huge increase in green space and trees transformed the atmosphere and contributed to the global goal of greenhouse gas reduction. Energy these days comes from renewable sources beyond the city. These vary from region to region and in the UK, for example, a gamechanger has been the inclusion of large-scale predictable tide power in the mix.

The high density and mixed use of the city, organized into nodes like this one, means you can do most of your shopping, get a lot of your entertainment, hook up with your football team and your local street run and get to work all without even getting on the metro. Green spaces like the one below your window where you can stroll, meet friends, cycle and just enjoy the environment are an integral part of the city. They link its different areas and services together, make the whole scene more social and communal, connecting people with each other and also reconnecting city-dwellers with nature. These days the city's inhabitants are much more concerned to protect their environment and the planet than in the concrete jungles of thirty years ago.

Your workplace is not much more than an easy stroll away, but you've chosen to jog. Along the way you pass many independent shops and businesses. The compact city has encouraged and incentivized a renaissance of localized and independent businesses and your neighbour has also explained that this has broken down some of the previous segregation of cities into prosperous areas and poorer ghettos. With decent work available in many different businesses, and greater availability of affordable housing the city is mixed and inclusive in income groups as well as uses.

Alongside you on the service road cargo e-bikes are shuttling to and fro, linking businesses and shipping supplies. In the distance is one of the node's business hubs where goods arrive in the area by train for local distribution. Not only in the ribbon of allotments beyond the service road but also on balconies, in backyards and on roofs stuff is growing everywhere. In some cases, this is food that's produced and sold commercially but lots of locals have got in on the act, growing their own food too.

This green boulevard has another function too. The city was, years ago, increasingly flooded during the extreme storms driven by climate change. The vast areas of concrete and tarmac magnified the problem. Now these green areas are designed to capture the surface water, and wells set into them divert it into groundwater.

You've seen documentaries about the state of cities like this thirty years ago. The way their centres were increasingly polluted, unhealthy and dangerous and how their unmanaged suburban sprawl was eating up valuable rural land and creating longer and longer gas-guzzling journeys for suburban commuters. And you've seen documentaries too about the bigger global picture. Cities like yours producing more than 60 per cent of global emissions, changing the climate and progressively destroying the planet – your home.

Those documentaries showed that there were huge transition costs involved in shaping what your city is now, and making the whole world a safer, healthier, less disaster-prone place. Those costs were not only economic. People didn't want to bear their own personal costs – giving up cars, giving up the suburban dream, giving up to a large extent meat-based food. But the crises of the 2020s gradually motivated the many disparate movements, campaigns and civil society organizations to move beyond their own agendas and ask how they could work together to escape the spiralling crisis. Industry pretended to act, but protected its own interests. Governments and international frameworks seemed incapable of meeting even modest targets for change. Several visionary leaders across Asia, Africa, Europe and Latin America gradually created the impetus for a citizen-led movement for change. Named, originally, 'Citizens for life' it somehow became 'Cityzens for life' as city-dwellers finally realized their urban habitats were becoming unliveable. Active citizens increasingly realized they could vote, not only with a tick on a political ballot but with their credit cards. The movement grew from local roots in towns and in many localities within cities, taking what action it could locally but uniting, as previous social campaigns had, with a growing voice. Eventually this citizen-led movement forced industry to invest in innovative solutions and products to offer within this new reality, and government stimulated change by adapting its incentives, subsidies and regulatory frameworks to take notice and start framing liveable futures.

Though you spend much of your life within the city, and much of that in a small radius around your home, you travel when you wish. High-speed trains take you to relatives in the north, and to holiday destinations in the south. You've travelled further afield too. The rail network links to national air hubs dedicated to international rather than internal travel. High-density aircraft link international hubs. Low-density first class travel, use of private jets and internal air travel are all things of the past, but you can, if you choose, see more of your world.

When you've taken such trips, you've seen that the unmanaged exploitation of the planet over the previous two centuries has left many scars, environmental and social. It will take many decades for those scars to begin to heal, but the most recent estimates say that the worst is behind you.

You've arrived at your workplace. The job's satisfying and the business is going well. More important in a way than all the technical achievements of the transition and than averting the climate catastrophe is that life is now – for many people, not only those with wealth and resources – much more healthy and satisfying than it was just three decades ago. As the voices of Cityzens for life grew in noise and urgency it was clear that they wanted a better, sustainable future, and compact cities were one of the models for achieving that. It was far from clear how that major transition would be achieved at the outset but the important thing was the vision and the goals. Somehow the inventiveness and innovation of people, of society – and in the end of industry and government – made it happen. As far as you're concerned the costs were worth it.

This speculative scenario reflects much thinking about how to make cities not only more technically sustainable but better places to live.[35] The different elements of compact cities are interconnected.[36] High-density housing and mixed activities within local areas, incorporating accommodation, business, education and other services, stimulate local trade, reduce travel and energy consumption, and promote social cohesion. Mixed use and increased local trade and business create employment opportunities and help to integrate previously segregated prosperous and poor populations. Green spaces contribute to the environmental and social qualities of the city as well as to disaster management. They also reconnect city populations with the living planet as an intrinsically valuable thing, the 'global commons' rather than just a resource to be exploited.

Substantial reduction in private road travel is an integral consequence of compact city design. While such travel is assumed to be largely electrified by 2050, driving down local emissions, the overall reduction drives down global energy consumption. This leaves the challenge of air travel. The concept of mixed mode travel in which short-haul movements are by rail linking long-haul travel by air has been shown to drive down emissions.[37] In addition the concept of high-density air travel for long-haul air travel can substantially drive down individual travellers' carbon footprint. A World Bank study used to assess its staff travel policies showed that the footprint for business class air travel was three times that of economy, for first class travel it was *nine* times. Increasing the passenger density on aircraft therefore dramatically reduces per capita emissions.[38]

Future cities: From disaster making to changemaking

While the driving force for much action on compact cities is specific concern about climate change, it is highly relevant to the goal of driving down risk creation – tackling both hazards and vulnerabilities – both of which we have seen increasing as we've charted the growth of cities. Table 3 summarizes elements of risk creation in cities, showing the hazards and vulnerabilities that resulted from them. We can see how this potential future might address both through a comparison table (Table 4) which suggests ways that a future city scenario would mitigate these risks.

Remember that this is a hypothetical scenario, a vision for an alternative to current unsustainable trends. On my first encounter I dismissed it as wishful thinking and it begs huge questions about how public concern, which is well documented, would lead to this kind of transformatory action. In addition, ideas like this are not universally welcomed. 'The fifteen-minute city' – a concept allied to that of the compact city[39] – has been the subject of conspiracy theories claiming that it's an attempt at state control, and that people will be restricted to local zones within cities. The UK Conservative Party even named this supposed threat as one they would fight against.[40] However, we've learnt already that media and political depictions of public attitudes are misleading, tending to amplify extreme views. In practice, incremental efforts towards compact cities have been welcomed. This scenario was based partly on case studies from Gothenburg and Helsingborg in Sweden, where public views are largely positive.[41] Another project in Los Angeles, initially driven by the need to manage increasing flooding, created green strips like those in the scenario which have also been welcomed.[42] In Valencia, Spain, diverting an entire river to reduce flooding led to creation of linear parks right through the city in the dry river bed, as an amenity for everyone.[43]

Table 4 Risk Creation and Mitigation in Cities

Cities 1760–2025: Risk Creation	Hazards and Vulnerabilities	Mitigation in the Cities of 2050
BUILT ENVIRONMENT		
Poor services, water, waste, sanitation, etc.	Magnifying health vulnerabilities	Integrated high-quality services
Informal housing: slums and shanties	Magnifying health and psychological vulnerabilities	Transition to formal housing in mixed-use mixed-economy localities
Overdeveloped built environment	Magnifying climatic hazards, i.e. flooding	Integrated sustainable built environment managing risks
Increasing road transport, infrastructure, pollution and emissions	Magnifying local health hazards and global climate hazards (also global impacts)	Local nodes, non-vehicle and public transport reducing and decarbonizing vehicle use
Lack of public transport and safe walking and cycle routes	Increased vehicle use, emissions and pollution hazards (also global impacts)	As above
Hazard-prone structures	Magnifying climate and geophysical hazards	Context-appropriate resilient building
LAND USE		
Urban and suburban sprawl	Magnifying climate and health hazards through increased vehicle use	Compact high-density cities reducing sprawl
Development on low-lying and unstable areas	Magnifying climate hazards, i.e. flooding	Integrated disaster-resilient land use design
Degrading natural environment and biodiversity	Magnifying ecological hazards (also global impacts)	Integration of built and natural environment
Lack of green space and tree cover	Increased physical and mental health hazards, decreased air quality	Recreate open spaces, tree planting
INDUSTRIAL ACTIVITY		
Exploitation of workers: low wages, insecurity	Magnifying economic vulnerability	Localized circular economies increasing decent work access
Hazardous working conditions	Magnifying health vulnerability	Social demand for socially responsible industry
Increasing materials consumption	Magnifying environmental hazards (also global effects)	Reducing pace of consumption, increasing recycling and reuse
Increasing pollution	Magnifying health and environmental hazards (also global effects)	Reduced pollution from localized circular decarbonized economies

Table 4 (continued)

Cities 1760–2025: Risk Creation	Hazards and Vulnerabilities	Mitigation in the Cities of 2050
Increasing emissions	Magnifying health and climatic hazards (also global effects)	Reduced pollution from localized circular decarbonized economies
SOCIAL IMPACT		
Inward migration from rural areas	Growth of health hazards in economically segregated informal housing	Investment in sustainable rural livelihoods; decent housing and work in cities
Lack of access to health services	Increased health and disease vulnerabilities	Provision of high-quality services in local nodes
Lack of access to education	Persistent poverty, crime	Universal access to education
Increasing segregation into poor/prosperous	Magnifying economic vulnerability	Mixed use rather than zoned design reducing social segregation
Social isolation	Magnifying psychosocial vulnerability, family fragmentation and breakdown	Increased social space, social interaction, physically and psychologically healthy environment; strengthened familial relationships
Weak links between CSO and local government	Inability to match services and risk reduction to local knowledge and capacities, magnifying hazards and vulnerabilities	Locally driven collaborations and action drawing on local knowledge and experience, claiming urban commons and strengthening CSO/ government relationships

Considering the social, economic and environmental realities of our current unsustainable trajectory it is clear that substantive change, whether reflecting this or another vision for a future we would choose, is necessary. This scenario hints at ways in which change can be driven locally by citizens and civil society groups and at larger scales by unifying movements and campaigns. It highlights the role of the public as *active citizens* rather than *passive consumers*. How could this shift from passive to active come about?

Chapter 13

CHANGEMAKING FOR THE FUTURE WE CHOOSE

How might the public act to drive what might seem impossibly large but vital changes in the way our cities and our world are organized? Starting small, at local level people are prepared to collaborate to protect their commons, as Ostrom showed in a range of case studies. The actions I witnessed among the GNDR membership give me confidence in this scale of action. One example was the launch of an 'Action at the Frontline' competition inviting people to tell stories of local level action. We were hoping to gather ten cases for presentation at a UNDRR disaster reduction conference. We were snowed under with fifty-nine contributions from local level groups telling stories of ways they'd taken action to reduce the effects of disasters in their localities. The winner was a group from Central America who'd solved a problem of water management and flooding by building a dam, based entirely on local voluntary efforts.[1] In fact, the energy of local groups often extends to them subverting projects imposed on them by external agencies, shifting their goals through local efforts towards what they see as priorities for action. People in all these cases are not passive, but active and knowledgeable.[2] It's quite a stretch, however, to move from these local collaborations to social demand on a scale that would achieve changes required in the future cities' scenario described above, for example. The VFL reports produced by GNDR had some influence on national government policies and on the tone and emphasis of the International Disaster Reduction framework, but the scale of action and influence we are considering here are much greater. In my twin roles as practitioner and researcher at GNDR it was important to me to learn from wider action and research how this combination of local and global campaigning could be conducted successfully. There's a wide range of experience and thinking on this topic.

The challenges of global changemaking

Examples of other campaigns show that in achieving their objectives they need to engage their public, define a change which is understood and desired, stimulate a change from passive consumerism to active citizenship and generate such a groundswell of public action that political and other actors start to listen. They face the challenges of lack of information and understanding, resistance by vested interests and also the conflicted state of civil society itself, operating in a marketplace in which individual organizations compete for attention and funding.

Movements which have surmounted these challenges and succeeded in creating societal transitions include, for example, the abolition of slavery, the establishment of trade unions, the anti-apartheid movement in South Africa and the American civil rights movement. These movements suggest change is possible, though they all contain elements of struggle over significant time periods.

In developing VFL and other projects at GNDR we drew on the experiences of some of these other civil society-led networks and campaigns. Two important sources were the book *Global Action Networks* by Steve Waddell[3] and the report 'Campaigning for International Justice' by Brendan Cox.[4] They considered a range of networks and campaigns. Some examples (from both sources) are:

- **Forestry Stewardship Council**: Aims to protect forests by certification of timber extracted in ways that maintain sustainability.
- **Transparency International**: Formed when the World Bank drew back from efforts to address corruption and a staff member left to establish a network and campaign to tackle this.
- **The International Campaign to Ban Landmines**: Formed to meet the objective of its title, deliberately keeping its focus narrow (not including cluster munitions, for example) in order to force a clear result.
- **Jubilee 2000** ('Drop the Debt'): Formed to campaign against the imposition of unjust debt repayments on developing world countries; focused on debt remission in specific countries.
- **Make Poverty History**: Called for Trade Justice, Dropping the Debt and More and Better Aid. Unlike the other examples listed the objectives were broad and general.

These campaigns have had varied success in reaching their objectives. Some focused on a particular 'moment' or specific change, others are ongoing. A common success factor is a clear and specific focus on a particular change which might iterate towards a larger objective. The outlier is 'Make Poverty History' which made broader demands. Cox argues that for this and organizational reasons it was relatively ineffective.

Learning from failure: The 'Global Campaign for Climate Action'

On re-reading the studies several years after we first made use of them, I was particularly struck by the case of the 'Global Campaign for Climate Action', active in the run-up to the chaotic COP15 in Copenhagen (2009). The case is instructive even though the action didn't achieve its objectives, in fact particularly *because* of that. Often the best learning comes from understanding failure. Based on interviews with observers and participants in the project Cox highlights the following:

- *Overly ambitious objectives* of securing a 40 per cent cut in emissions in industrialized countries from a 1990 baseline, a legally binding agreement, and large-scale adaptation and mitigation financing, which were seen at the time as 'out of the ballpark'. With the benefit of hindsight, we see that these objectives are still being sought, with very limited success, at COP28.
- *Focusing efforts on a large-scale international event*, rather than building from national level.
- *Political naivety*, ignoring the role of emerging economies, resulting in even greater resistance from industrialized nations.
- *Negative messaging*, focusing on an apocalyptic future rather than the opportunities of a low-carbon economy.
- *Narrow messaging*, the campaign was dominated by environmental and development groups and its messaging was therefore technical rather than mainstream.
- *Short time scale*, the campaign was only really active for several months leading up to COP15.
- *Incoherent structure and identity:* The network used a 'flotilla' structure in which member organizations were loosely organized around the campaign identity and messages, but kept their own branding. This weakened the visibility and focus of the campaign.

Most participants and observers felt the campaign failed to achieve impact for the above reasons, and though again establishing causality is difficult, the campaign's claim that it created momentum for change is contradicted by events, as the COP process fell into the doldrums for several years following. Turning the failings of this campaign on its head, what would an effective campaign look like?

Elements of effective campaigns

I'm inspired by hopeful, positive visions for a better future, for example, in a piece from National Geographic which asked activist individuals for their future visions. Many of these were constructive, positive and focused on people as well as the planet.[5] This and other characteristics appear from the evidence of the above studies as keys to effective change. The most important elements are summarized below:

Start local and build: A large-scale citizen movement needs to start with people, and then draw together a wider coalition. This was the case, for example, with the successful Landmines and Jubilee 2000 campaigns.

Present positive visions for the future: We've seen from our earlier examination of public attitudes and behaviour that people engage with positive visions rather than negative messages.

Take an incremental approach: Rather than large-scale unspecific goals, fashion specific achievable 'asks', which build iteratively towards the bigger vision (as several of the campaigns highlighted above have).

Build a coalition with a strong and clear identity: Effective coalitions have had clear leadership, identity and messages rather than being compromised by a 'flotilla' approach as with the 'Global Campaign for Climate Action' or being compromised by the demands of large member NGOs – as the 'Make Poverty History' campaign was, according to Cox.

Combine strong leadership with strong ownership of members: We learned this the hard way at GNDR, and the Landmines campaign (and others) combine strong national movements with clear central leadership.

Take time: Change takes time and large-scale change takes even longer. The large-scale social transformations we mentioned took decades.

Changemaking for the future we choose

We've seen that the public, while key to finding pathways to a sustainable future, are often paradoxical in their attitudes, recognizing the need for change but feeling powerless to achieve it. Taking the core message that people respond much more strongly to positive visions than to doom-laden, apocalyptic concerns we've taken the example of a particular scenario for the transformation of cities, which is where a growing majority of us live. We've seen that this vision encompasses many dimensions of risk creation. It's an illustrative example of a positive vision for a future which drives down the magnification of hazards and vulnerability. We've considered a range of campaigns and networks designed to mobilize the public in changemaking, and distilled from them elements of effective campaigns. The scale of change required to transform our society and our planet towards a safe, sustainable and prosperous future is larger than that in any of the social movements and campaigns we have considered, and yet the consequences of not moving towards this transformation are large scale and disastrous. Somehow, out of the energy of many disparate movements it seems necessary for a coalition to emerge, growing from the ground up, clearly focused, prepared to set iterative goals for change over a period likely to span decades. Simple really!

CONCLUSION: UNNATURAL DISASTERS, CONVENIENT UNTRUTHS

I referred to Ai Weiwei's angry writing about the Sichuan disaster in the preface:

> The disaster makers always go free and the innocent are punished.[1]

I asked, in response to Weiwei's words:

> Who, or what, are the disaster makers? Is disaster risk created by malignant individuals or organisations, as the term suggests? Or are wider and historic forces in play? Is the creation of ever-increasing risk the result of a wider system? How can this growing disaster risk be addressed?

I suggested that the answer is there in the book's subtitle: *Tackling Unmanaged Growth for Sustainable Futures*, but that the story is more complicated than that brief strapline suggests.

Al Gore's 2006 film *An Inconvenient Truth* presented unavoidable facts about climate change and its consequences.[2] Nearly two decades later the Keeling curve charting CO_2 emissions – an early influence on Gore when he encountered it in the 1960s – still trends inexorably upwards. Greenhouse gas emissions are just one cause of unnatural disasters, perhaps currently the most measured and publicized, but risk is being created much more widely than just through fossil fuel consumption. Our cities are crumbling under the weight of unmanaged growth leading to unsustainable and inequitable overdevelopment. Beyond the cities, fields and forests, rivers and seas are overexploited, polluted and poisoned. The vital interlocking system of plants, insects and animals which sustain their life and productivity is collapsing.

It's not enough to try to protect current patterns of development, wrapping them around with disaster reduction activities, as disasters are embedded in development itself. As UNDRR's 2023 review reported despairingly: 'Despite commitments … to pursue sustainable and regenerative development, current societal, political and economic choices are doing the reverse. Intensive and extensive risks are growing at an unprecedented rate.'[3]

To invert Gore's title, *convenient untruths* allow this continuing contradiction – cognitive dissonance – between commitments to sustainable development and continuing creation of risk. They are embedded deep in the system, in our ideologies, in our psyches. They include the following.

Convenient untruths

1. **Cities and the planet (our urban and global commons) are infinitely exploitable and degradable resources**. An untruth that ignores the damage we've done, the damage we're doing, the trend-lines for the social, economic and environmental degradation of our home over the next generation and the failure to recognize the value of our local, urban and global commons.
2. **Free markets work**. The idea of self-managing free markets is a deeply held untruth today. We have seen that in reality industry, through concentration, increasing power, branding, marketing and lobbying, protects itself from the pressures of free markets to pursue self-preservation and profit rather than the wider interests of 'consumers'.
3. **Rejecting regulation.** Though mainstream economics recognizes the necessity for rules and regulation to protect against market failures and externalities, resistance to this and rejection of 'red tape' is a deeply held untruth.
4. **Consumers not citizens.** Industry, government and media depict the public as passive, playing their role in the disaster making system as consumers. This untruth persuades the public they are unable to take their part in securing a better future as active citizens.
5. **The present is as good as it gets**. Industry and government are united in this untruth, as change comes with transition costs which are deeply uncomfortable for them. For the public and the planet

there has to be a better future as persisting in the present trajectory will lead to a bleak future.

Ai Weiwei's words suggested the malevolent self-interested actions of disaster makers. I conclude that while individual and organizational disaster makers lie behind particular disaster events, leading, for example, to the human cost of the Sichuan earthquake, these are part of a wider systemic problem. Convenient untruths underpin an ideology sustaining a disaster making system, one which is now increasingly unsustainable as its consequences exceed planetary boundaries.

The UN system is capable of *recognizing* the problem – 'The Sustainable Development Goals are disappearing in the rear-view mirror, as are the hopes and rights of current and future generations.'[4] But it is clearly failing in its mission to change course.

Maybe because I was involved myself in a change-making network and maybe also because (as Gore does) I think of social changes that have been achieved historically, I believe change-making can come from the public. I've drawn on studies and survey data suggesting that passive consumer roles are shaped by industry and the media, that people are actually concerned but unclear how change can come about. Change stories which would inspire and motivate the public start, I believe, not with any of the spectrum of ideas about the *means* of change, from techno-fixes to revolution. The starting point is through engaging the public with clear *visions* for a better future, gradually uniting to press for change with their voices, their voting slips and their credit cards.

If we embrace our role as active citizens, gather around our visions for futures we would choose and act locally and unite globally to create a system that works for all of us within our global commons – our planetary home – we might address the root causes of unnatural disasters and turn from unmanaged growth to sustainable futures.

NOTES

Preface: A personal disaster journey

1 A psychology student who'd left Ukraine temporarily during the 2022 war drew my attention to the concept of 'collective trauma' which she was studying in her country's context but which she suggested was relevant to the pandemic experience: Thomas Hübl, *Healing Collective Trauma: A Process for Integrating Our Intergenerational & Cultural Wounds* (Boulder, CO: Sounds True, 2020). See also: Roxane Cohen Silver, E. Alison Holman and Dana Rose Garfin, 'Coping with Cascading Collective Traumas in the United States', *Nature Human Behaviour* 5, no. 1 (January 2021): 4–6, https://doi.org/10.1038/s41562-020-00981-x.

2 'Public Support Majority of Net Zero Policies ... Unless There Is a Personal Cost', Ipsos, 18 October 2021, https://www.ipsos.com/en-uk/public-support-majority-net-zero-policies-unless-there-is-a-personal-cost; United States: Alec Tyson, Cary Funk and Brian Kennedy, 'What the Data Says about Americans' Views of Climate Change', *Pew Research Center* (blog), accessed 22 May 2023, https://www.pewresearch.org/short-reads/2023/04/18/for-earth-day-key-facts-about-americans-views-of-climate-change-and-renewable-energy/; EU: European Commission. Directorate General for Climate Action, *Climate Change: Report* (LU: Publications Office, 2021), https://data.europa.eu/doi/10.2834/437.

3 Michael Wall and Ian Davis, 'Christian Perspectives on Disaster Management: A Training Manual' (Interchurch Relief and Development Alliance, 1992), https://desastres.unanleon.edu.ni/pdf/2002/agosto/PDF/ENG/DOC3788/DOC3788.HTM.

4 Ian Davis, *Shelter After Disaster* (Oxford: Oxford Polytechnic Press, 1978), http://shelterprojects.org/other/siteplanning/site-planning-references/Davis-1978-shelter%20after%20disaster.pdf.

5 *When It Comes to the Crunch*. VHS (Tearfund, 1993).

6 Piers Blaikie, Terry Cannon, Ian Davis and Ben Wisner, *At Risk: Natural Hazards, People's Vulnerability and Disasters*, 2nd edn (London: Routledge, 2014), https://doi.org/10.4324/9780203714775.

7 *On Solid Ground* (Tearfund, 2003).

8 'Views from the Frontline | Projects at GNDR', GNDR, accessed 31 July 2023, https://www.gndr.org/project/views-from-the-frontline/.

9 Ai Weiwei, 'Our Duty Is to Remember Sichuan', *The Guardian*, 25 May
   2009, sec. Opinion, https://www.theguardian.com/commentisfree/2009/
   may/25/china-earthquake-cover-up.

Introduction

1 'Half of the Global Population Lives on Less Than US$6.85 per Person per
   Day', 8 December 2022, https://blogs.worldbank.org/developmenttalk/
   half-global-population-lives-less-us685-person-day. The report notes that
   the most widely used measure is of 'extreme' poverty in poor countries
   (currently $2.15 per day), but that 70 per cent of the world's population live
   in 'middle income' countries where $6.85 per day (an annual income of
   $2,500 or less) is a more meaningful poverty line.

Chapter 1

1 Field visit 15 June 2001. Lydia was recorded as part of a series of video
   programmes capturing local-level experiences of response and recovery
   after Hurricane Mitch. *On Solid Ground*. VHS. Tearfund. 2003.
2 William C. Smith, 'Hurricane Mitch and Honduras: An Illustration
   of Population Vulnerability', *International Journal of Health
   System and Disaster Management* 1, no. 1 (2013): 54, https://doi.
   org/10.4103/2347-9019.122460.
3 Norman Kerle, Benjamin Van Wyk De Vries and Clive Oppenheimer, 'New
   Insight into the Factors Leading to the 1998 Flank Collapse and Lahar
   Disaster at Casita Volcano, Nicaragua', *Bulletin of Volcanology* 65, no. 5
   (July 2003): 331–45, https://doi.org/10.1007/s00445-002-0263-9.
4 Phil Gunson and By Filadelfo Aleman, 'Hundreds Perish in Nicaraguan
   Mudslide', *The Guardian*, 2 November 1998, sec. Environment, https://
   www.theguardian.com/environment/1998/nov/02/weather.climatechange.
5 Field visit, North-east Japan. 19 March 2015.
6 Akemi Ishigaki, Hikari Higashi, Takako Sakamoto and Shigeki Shibahara,
   'The Great East-Japan Earthquake and Devastating Tsunami: An Update
   and Lessons from the Past Great Earthquakes in Japan since 1923', *The
   Tohoku Journal of Experimental Medicine* 229, no. 4 (2013): 287–99, https://
   doi.org/10.1620/tjem.229.287.
7 Field visits Kathmandu, Nepal, 11 April 2014 and 1 November 2017.
8 H. Hazarika, N. P. Bhandary, Y. Kajita, K. Kasama, K. Tsukahara and
   R. K. Pokharel, 'The 2015 Nepal Gorkha Earthquake: An Overview of
   the Damage, Lessons Learned and Challenges', *Lowland Technology
   International* 18, no. 2 (2016): 105–18.

9 Gabriel Scally, Bobbie Jacobson and Kamran Abbasi, 'The UK's Public Health Response to Covid-19', *BMJ* 369 (15 May 2020): m1932, https://doi.org/10.1136/bmj.m1932.

10 'COVID-19 Map', Johns Hopkins Coronavirus Resource Center, accessed 19 March 2024, https://coronavirus.jhu.edu/map.html.

11 Stephanie C. Martin, 'Past Eruptions and Future Predictions: Analyzing Ancient Responses to Mount Vesuvius for Use in Modern Risk Management', *Journal of Volcanology and Geothermal Research* 396 (May 2020): 106851, https://doi.org/10.1016/j.jvolgeores.2020.106851.

12 'Pompeii in Pictures', accessed 24 May 2023. https://www.pompeiiinpictures.com/pompeiiinpictures/R0/Casts.htm.

13 'Presidency of the Republic of Turkey: President Erdoğan in Quake-Hit Şanlıurfa', accessed 24 May 2023, https://www.tccb.gov.tr/en/news/542/142899/president-erdogan-in-quake-hit-sanliurfa.

14 UNISDR, ed., *Making Development Sustainable: The Future of Disaster Risk Management*, Global Assessment Report on Disaster Risk Reduction, 4.2015 (Geneva: United Nations, 2015), p. 25.

15 Terry Gibson, 'Learning How to Build Back Better', *Monday Developments: Interaction USA* (June 2010).

16 Peter Beaumont, 'Turkey Earthquake Death Toll Prompts Questions over Building Standards', *The Guardian*, 7 February 2023, sec. Global development, https://www.theguardian.com/global-development/2023/feb/07/turkey-earthquakes-death-toll-prompts-questions-over-building-standards.

17 Stephanie C. Martin, 'Past Eruptions and Future Predictions: Analyzing Ancient Responses to Mount Vesuvius for Use in Modern Risk Management', *Journal of Volcanology and Geothermal Research* 396 (1 May 2020): 106851, https://doi.org/10.1016/j.jvolgeores.2020.106851.

18 UK National Audit Office, 'The Government's Preparedness for the COVID-19 Pandemic: Lessons for Government on Risk Management', 15 November 2021, https://www.nao.org.uk/wp-content/uploads/2021/11/The-governments-preparedness-for-the-COVID-19-pandemic-lessons-for-government-on-risk-management.pdf.

19 'World Health Organization Fears "Disease X" Could Cause a Global Pandemic', *The Independent*, 11 March 2018, https://www.independent.co.uk/news/science/disease-x-what-is-infection-virus-world-health-organisation-warning-ebola-zika-sars-a8250766.html.

20 Global Preparedness Monitoring Board, 'A World at Risk', Annual Report, September 2019, https://www.gpmb.org/annual-reports/annual-report-2019.

21 'How the 4 Biggest Outbreaks since the Start of This Century Shattered Some Long-Standing Myths', accessed 26 May 2023, https://www.who.int/news/item/01-09-2015-how-the-4-biggest-outbreaks-since-the-start-of-this-century-shattered-some-long-standing-myths.

22 The reports were all made available by a transparency initiative: Moosa, 'Exclusive – Seven Secret Pandemic Reports (Including Alice)', *CygnusReports.Org* (blog), 8 October 2021, https://cygnusreports.org/seven-reports/.

23 Telegraph Reporters, 'Boris Johnson "Scrapped Cabinet Pandemic Committee Six Months before Coronavirus Hit UK"', *The Telegraph*, 13 June 2020, https://www.telegraph.co.uk/politics/2020/06/13/boris-johnson-scrapped-cabinet-pandemic-committee-six-months/.

24 Harry Davies, David Pegg and Felicity Lawrence, 'Revealed: Value of UK Pandemic Stockpile Fell by 40% in Six Years', *The Guardian*, 12 April 2020, sec. World news, https://www.theguardian.com/world/2020/apr/12/revealed-value-of-uk-pandemic-stockpile-fell-by-40-in-six-years.

25 AFP, 'Deforestation Worsened Impact of Hurricane Mitch', November 1998, https://wgbis.ces.iisc.ac.in/envis/doc98html/biodfore1124.html.

26 'Nicaragua Forest Information and Data', accessed 26 May 2023, https://rainforests.mongabay.com/deforestation/2000/Nicaragua.htm.

27 United States Geological Survey, 'Socioeconomic and Environmental Impacts of Landslides', July 2001, https://pubs.usgs.gov/of/2001/ofr-01-0276/.

28 Akemi Ishigaki, Hikari Higashi, Takako Sakamoto and Shigeki Shibahara, 'The Great East-Japan Earthquake and Devastating Tsunami: An Update and Lessons from the Past Great Earthquakes in Japan since 1923', *The Tohoku Journal of Experimental Medicine* 229, no. 4 (April 2013): 287–99, https://doi.org/10.1620/tjem.229.287.

29 Pers. Comm. during field visit Kathmandu, Nepal, 1 November 2017 with staff of NSET. See also: 'Why Was Nepal Badly Prepared for the Earthquake on 25 April 2015? › Friedrich-Alexander-Universität Erlangen-Nürnberg', accessed 5 June 2023, https://www.fau.eu/2015/05/21/news/why-was-nepal-badly-prepared-for-the-earthquake-on-25-april-2015/ and Alexandra Titz, 'Geographies of Doing Nothing – Internal Displacement and Practices of Post-Disaster Recovery in Urban Areas of the Kathmandu Valley, Nepal', *Social Sciences* 10 (22 March 2021): 110, https://doi.org/10.3390/socsci10030110.

30 Benedetto De Vivo and Giuseppe Rolandi, 'Vesuvius: Volcanic Hazard and Civil Defense', *Rendiconti Lincei* 24, no. 1 (March 2013): 39–45, https://doi.org/10.1007/s12210-012-0212-2.

31 'Disasters', in *Concise Oxford English Dictionary* (Oxford University Press, 2002).

32 UNISDR, *Living with Risk: A Global Review of Disaster Reduction Initiatives*, 2004 version (New York: United Nations, 2004).

33 'Sendai Framework Terminology on Disaster Risk Reduction | UNDRR', 9 March 2023, http://www.undrr.org/terminology.

34 GNDR, 'Views from Frontline Report 2009 | Resources at GNDR'
    (GNDR, 2009), https://www.gndr.org/resource/views-from-the-frontline/
    views-from-the-frontline-report-2009/.

35 Field visit to Uhambingeto, Tanzania, May 2008.

36 Field visit to Cocotomey, Cotonou, Benin, March 2009.

37 Field visit to Mahminet region, Eritrea, March 2003.

38 Field visit to Limbe, Cameroon, September 2011.
    'Views from the Frontline' survey data accessed
    at: https://app.powerbi.com/view?r=eyJrIjoiOTc2ZmYwND
    ktMWJkMS00YTg3LWE2NWQtNGMxZ%20
    DJjNmIxMWI1IiwidCI6IjZjYTY0NjFmLWI1MjItNDc2MC05NTcwL
    TEyZGFmNWU4ZDgxNSIsImMiOjh9. Accessed 5 June 2023.

39 The concept of attritional risk and the substantial losses from this risk
    type are well recognized in the insurance industry. See, for example,
    Luke Gallin, 'Lower Cats, but Elevated Attritional Losses to Hit P&C Re/
    Insurers in Q1: Analysts – Reinsurance News', ReinsuranceNe.ws, 8 April
    2019, http://www.reinsurancene.ws/lower-cats-but-elevated-attritional-los
    ses-to-hit-pc-re-insurers-in-q1-analysts/.

40 These accounts are included in a special edition of the *International
    Journal of Disaster Prevention and Management*: John Norton and Terry
    David Gibson, 'Introduction to Disaster Prevention: Doing It Differently
    by Rethinking the Nature of Knowledge and Learning', *Disaster Prevention
    and Management: An International Journal* 28, no. 1 (1 January 2019): 2–5,
    https://doi.org/10.1108/DPM-02-2019-323. The material can also be
    accessed at www.drr2dev.com.

41 See, for example, the UNDRR assessment which states 'Risk creation is
    outstripping risk reduction'. UNDRR, 'GAR2022: Our World at Risk', 2022,
    https://www.undrr.org/gar2022-our-world-risk-gar.

42 Sarwar Bari, 'Waiting for Politics at the Mercy of River: Case Study of
    an Enduring Community', *Disaster Prevention and Management: An
    International Journal* 28, no. 1 (1 January 2018): 42–9, https://doi.
    org/10.1108/DPM-06-2018-0187.

43 H. Kit Miyamoto and Tom Chan, 'Assessment of Damage and Lessons
    Learned', *Structure Magazine*, January 2009, https://www.structuremag.
    org/wp-content/uploads/2014/08/C-Str-Perf_China_Earthquake-Miyam
    oto-Gilani-Chan-Jan-09.pdf.

44 Ai Weiwei, 'Our Duty Is to Remember Sichuan', *The Guardian*, 25 May
    2009, sec. Opinion, https://www.theguardian.com/commentisfree/2009/
    may/25/china-earthquake-cover-up.

Chapter 2

1   Stuart Heritage, 'Matt Hancock's Vaccine Rollout Was Inspired by
    Contagion. Here's What He Should Watch Next', *The Guardian*, 4 February
    2021, sec. Film, https://www.theguardian.com/film/filmblog/2021/feb/04/
    contagion-film-matt-hancock-covid-vaccine-policy-hollywood.

2   Steve Jennings, 'Time's Bitter Flood: Trends in the Number of Reported
    Natural Disasters' (OXFAM, 1 May 2011), https://www.researchgate.net/
    publication/313847052_Time's_Bitter_Flood_Trends_in_the_number_of
    _reported_natural_disasters.

3   World Bank, 'Building Resilience', 18 November 2013, https://documen
    ts1.worldbank.org/curated/en/762871468148506173/pdf/826480WP0v10
    Bu0130Box37986200OUO090.pdf.

4   United Nations, 'International Decade of Natural Disaster Reduction
    Proceedings' (UNDRR, 9 July 1999), https://www.preventionweb.net/
    files/31468_programmeforumproceedings.pdf.

5   Tim Merrill, *Honduras: A Country Study* (Washington, DC: Library of
    Congress, 1993), https://babel.hathitrust.org/cgi/pt?id=mdp.39015031871
    166&view=1up&seq=159.

6   USAID, 'USAID_Land_Tenure_Honduras_Profile_0.Pdf' (USAID, 2011),
    https://www.land-links.org/wp-content/uploads/2016/09/USAID_Lan
    d_Tenure_Honduras_Profile_0.pdf.

7   Field visits Honduras September 2006, February 2015 and pers. comm.
    with residents.

8   Pers. comm. Nueva Suyapa, September 2006.

9   The following paper includes description of townships similar and
    adjacent to Nueva Suyapa: Laura E. R. Peters, Aaron Clark-Ginsberg,
    Bernard McCaul, Gabriela Cáceres, Ana Luisa Nuñez, Jay Balagna,
    Alejandra López, Sonny S. Patel, Ronak B. Patel and Jamon Van Den
    Hoek, 'Informality, Violence, and Disaster Risks: Coproducing Inclusive
    Early Warning and Response Systems in Urban Informal Settlements in
    Honduras', *Frontiers in Climate* 4 (2022), https://www.frontiersin.org/artic
    les/10.3389/fclim.2022.937244.

10 *Santa Rosa de Aguan*, DVD (Tearfund, 2002).

11 David Griffith, 'Migration, Labor Scarcity, and Deforestation in Honduran
    Cattle Country', *Journal of Ecological Anthropology* 18, no. 1 (December
    2016), https://doi.org/10.5038/2162-4593.18.1.3.

12 Field visit to Mosquitia region and Krausirpe, February 1990. *This Is Your
    Land*, VHS (Tearfund, 2001). Paul House, 'Forest Farmers: A Case Study
    of Traditional Shifting Cultivation in Honduras', *Network Paper-Rural
    Development Forestry Network …*, 1 January 1997, https://www.academia.
    edu/666608/Forest_Farmers_A_case_study_of_traditional_shifting_cult
    ivation_in_Honduras.

13 Frederick C. Cuny, *Disaster and Development Cuny* (New York: Oxford University Press, 1983).

14 E. L. Quarantelli, 'Disaster Studies: An Analysis of the Social Historical Factors Affecting the Development of Research in the Area', *International Journal of Mass Emergencies & Disasters* 5, no. 3 (1 November 1987): 285–310, https://doi.org/10.1177/028072708700500306.

15 'Timeline: Classic DRR Publications | PreventionWeb', 26 May 2022, https://www.preventionweb.net/timeline-classic-drr-publications.

16 'Citizens Disaster Response Centre', Citizens Disaster Response Center | CDRC, accessed 24 January 2024, https://www.cdrc-phil.com/.

17 'All India Disaster Mitigation Institute', The All India Disaster Mitigation Institute, accessed 24 January 2024, https://aidmi.org/.

18 Allan Lavell, Alonso Brenes and Pascal Girot, 'The Role of LA RED in Disaster Risk Management in Latin America – UNESCO Digital Library', in *World Social Science Report 2013*, 429–33, accessed 28 July 2023, https://unesdoc.unesco.org/ark:/48223/pf0000260479.

19 Phil O'Keefe, Ken Westgate and Ben Wisner, 'Taking the Naturalness Out of Natural Disasters', *Nature* 260 (15 April 1976): 566–67, https://doi.org/10.1038/260566a0.

20 Davis, *Shelter after Disaster*.

21 Paul Susman, Phil O'Keefe and Ben Wisner, 'Global Disasters: A Radical Interpretation', *Interpretations of Calamity* (1 January 1983): 263–83.

22 United Nations, 'International Decade of Natural Disaster Reduction Proceedings'.

23 *DRR Early Days – Ken Westgate* (UNDRR, 2022), https://www.youtube.com/watch?v=EFf8RYwMc64.

24 'Time to Say Goodbye to "Natural" Disasters', accessed 2 June 2023, https://www.preventionweb.net/blog/time-say-goodbye-natural-disasters.

25 Blaikie et al., *At Risk*, 2014.

26 Blaikie, Piers, Terry Cannon, Ian Davis and Ben Wisner. *At Risk: Natural Hazards, People's Vulnerability and Disasters*, 2nd edn (London: Routledge, 2014), p. 87 *et seq.*

27 Piers Blaikie, Terry Cannon, Ian Davis and Ben Wisner, *At Risk: Natural Hazards, People's Vulnerability and Disasters*, 2nd edn (London: Routledge, 2014), https://doi.org/10.4324/9780203714775.

Chapter 3

1 Susan C. Stonich, *I Am Destroying the Land! The Political Ecology of Poverty and Environmental Destruction in Honduras* (New York: Routledge, 2021), https://doi.org/10.4324/9780429041068.

2 Ibid.

3 *Santa Rosa de Aguan.*

4   Griffith, 'Migration, Labor Scarcity, and Deforestation in Honduran Cattle Country'.

5   *This Is Your Land.*

6   Peters et al., 'Informality, Violence, and Disaster Risks'.

7   Blaikie et al., *At Risk.*

8   Ian Burton, 'Forensic Disaster Investigations in Depth: A New Case Study Model', *Environment: Science and Policy for Sustainable Development* 52, no. 5 (31 August 2010): 36–41, https://doi.org/10.1080/00139157.2010.507144.

9   A. Oliver-Smith, Irasema Alcántara-Ayala, Ian Burton and Allan Lavell, *Forensic Investigations of Disasters (FORIN): A Conceptual Framework and Guide to Research* (IRDR, 2016), https://www.researchgate.net/publication/291349173_Forensic_Investigations_of_Disasters_FORIN_a_conceptual_framework_and_guide_to_research.

10 Ibid.

11 'Honduras: Assessment of the Damage Caused by Hurricane Mitch, 1998' (United Nations Economic Commission for Latin America and the Caribbean, 14 April 1999), https://repositorio.cepal.org/bitstream/handle/11362/25506/LCmexL367eng_en.pdf?sequence=1&isAllowed=y.

12 Ibid. William Smith, 'Hurricane Mitch and Honduras: An Illustration of Population Vulnerability', *International Journal of Health System and Disaster Management* 1, no. 1 (2013): 54, https://doi.org/10.4103/2347-9019.122460.

13 Pers. comm. September 2006, Nueva Suyapa.

14 *Santa Rosa de Aguan.*

15 Griffith, 'Migration, Labor Scarcity, and Deforestation in Honduran Cattle Country'.

16 Susman et al., 'Global Disasters'.

17 Kendra McSweeney and Oliver T. Coomes, 'Climate-Related Disaster Opens a Window of Opportunity for Rural Poor in Northeastern Honduras', *Proceedings of the National Academy of Sciences* 108, no. 13 (29 March 2011): 5203–8, https://doi.org/10.1073/pnas.1014123108.

18 A phrase popularized by Bill Clinton in work on response to the 2004 tsunami: William Clinton, 'Key Propositions for Building Back Better' (United Nations, 2006), https://www.preventionweb.net/files/2054_VL108301.pdf.

19 John Telford, Margaret Arnold and Alberto Harth, 'Learning Lessons from Disaster Recovery' (The World Bank, 2004), https://documents.worldbank.org/pt/publication/documents-reports/documentdetail/966021468034504310/learning-lessons-from-disaster-recovery-the-case-of-honduras.

20 Pierre Frühling, 'Turning Disasters into Opportunities' (SIDA, May 2002), https://cdn.sida.se/publications/files/sida2152en-turning-disasters-into-opportunities.pdf.

21 Ibid.

22 Telford et al., 'Learning Lessons from Disaster Recovery'.

23 Malcolm Rogers, 'What Happened to Honduras after Hurricane Mitch' (Christian Aid, 1999), http://web.archive.org/web/20070606090147/http://www.christianaid.org.uk/indepth/9910inde/indebt2.htm.

24 Telford et al., 'Learning Lessons from Disaster Recovery'; Frühling, 'Turning Disasters into Opportunities'.

25 Telford et al., 'Learning Lessons from Disaster Recovery'.

26 Ibid.

27 Ibid.

28 Ibid.

29 *Santa Rosa de Aguan.*

30 Griffith, 'Migration, Labor Scarcity, and Deforestation in Honduran Cattle Country'.

31 Telford et al., 'Learning Lessons from Disaster Recovery'.

32 'Honduras Crime Rate & Statistics 1990–2023', accessed 31 July 2023, https://www.macrotrends.net/countries/HND/honduras/crime-rate-statistics.

33 BBC, 'Honduras Struggles 10 Years after Mitch', 30 October 2008, http://news.bbc.co.uk/1/hi/world/americas/7682412.stm.

34 Ibid.

35 Sonia Nazario, 'Enrique's Journey | A Six-Part Times Series', *Los Angeles Times*, 16 July 2014, https://www.latimes.com/nation/immigration/la-fg-enriques-journey-sg-storygallery.html.

36 Ginger Thompson and Nazila Fathi, 'For Honduras and Iran, World's Aid Evaporated – The New York Times', *New York Times*, 2005, https://www.nytimes.com/2005/01/11/world/worldspecial4/for-honduras-and-iran-worlds-aid-evaporated.html.

37 'Everyday Disasters and Everyday Heroes: How Frontline Finds Out from Local People What Threats They Face | PreventionWeb', 9 March 2015, https://www.preventionweb.net/publication/everyday-disasters-and-everyday-heroes-how-frontline-finds-out-local-people-what.

38 Terry Gibson and GNDR, 'Frontline: From Local Information to Local Resilience', 2017, https://www.gndr.org/wp-content/uploads/2021/11/FRONTLINE_PUBLICATION-English.pdf.

39 GNDR, 'Views from the Frontline Report 2013' (GNDR, 2013), https://www.gndr.org/wp-content/uploads/2021/11/View-from-the-Frontline-Report-2015-English-Version.pdf.

40 UNISDR, ed., *GAR 2015 Making Development Sustainable: The Future of Disaster Risk Management*, Global Assessment Report on Disaster Risk Reduction, April 2015 (Geneva: United Nations, 2015), https://www.preventionweb.net/english/hyogo/gar/2015/en/home/. (Chapter 4: Extensive Risk).

41 Daniel M. Sabet, 'When Corruption Funds the Political System: A Case Study of Honduras' (Woodrow Wilson Center, 2020), https://www.

wilsoncenter.org/publication/when-corruption-funds-political-sys
tem-case-study-honduras.

42 Vinicius Madureira, 'Honduras Sentences Ex-Official to over 10 Years
    for Buying Useless Mobile Hospitals', Organised Crime and Corruption
    Reporting Project, 2022, https://www.occrp.org/en/daily/16455-hondu
    ras-sentences-ex-official-to-over-10-years-for-buying-useless-mobile-
    hospitals.

43 'Our History | Association for a More Just Society', accessed 31 July 2023,
    https://www.asj-us.org/who-we-are/history.

44 Anthony Oliver-Smith, Anthony, Irasema Alcántara-Ayala, Ian Burton,
    and Allan Lavell, 'The Social Construction of Disaster Risk: Seeking Root
    Causes', *International Journal of Disaster Risk Reduction* 22 (2017): 469–74,
    https://doi.org/10.1016/j.ijdrr.2016.10.006.

45 LA RED was a convening network for much of this early work. See: Lavell,
    Brenes and Girot, 'The Role of LA RED in Disaster Risk Management in
    Latin America – UNESCO Digital Library'.

46 Oliver-Smith et al., 'The Social Construction of Disaster Risk'.

Chapter 4

1 Anthony Oliver-Smith, Irasema Alcántara-Ayala, Ian Burton and Allan
    Lavell, 'The Social Construction of Disaster Risk: Seeking Root Causes',
    International Journal of Disaster Risk Reduction 22 (June 2017): 469–74,
    https://doi.org/10.1016/j.ijdrr.2016.10.006. (See diagram p. 471.)

2 Ulrich Beck, *Risk Society* (London: SAGE, 1992).

3 Blaikie et al., *At Risk*.

4 Gautam Mukunda, 'The Social and Political Costs of the Financial Crisis,
    10 Years Later', *Harvard Business Review*, 25 September 2018, https://
    hbr.org/2018/09/the-social-and-political-costs-of-the-financial-cri
    sis-10-years-later.

5 'The Economics of Climate Change' (Swiss Reinsurance, April 2021),
    https://www.swissre.com/institute/research/topics-and-risk-dialogues/
    climate-and-natural-catastrophe-risk/expertise-publication-econom
    ics-of-climate-change.html.

6 'Presidency of the Republic of Turkey: President Erdoğan in Quake-Hit
    Şanlıurfa'.

7 'Adapted from: Hazard | Understanding Disaster Risk', 9 June 2021, https://
    www.preventionweb.net/understanding-disaster-risk/component-risk/
    hazard.

8 Ishigaki et al., 'The Great East-Japan Earthquake and Devastating
    Tsunami', April 2013.

9 Patrick W. Keys, Victor Galaz, Michelle Dyer, Nathanial Matthews, Carl
    Folke, Magnus Nyström and Sarah E. Cornell, 'Anthropocene Risk', *Nature*

    Sustainability 2, no. 8 (22 July 2019): 667–73, https://doi.org/10.1038/s41
893-019-0327-x; Meera Subramanian, 'Anthropocene Now: Influential
Panel Votes to Recognize Earth's New Epoch', *Nature*, 21 May 2019,
https://doi.org/10.1038/d41586-019-01641-5.

10 UNISDR, *Making Development Sustainable*, p. 203.

11 'GAR 2022 Summary for Policymakers', accessed 1 December 2023,
https://www.undrr.org/media/79594/download?startDownload=true.

12 UNISDR, *Making Development Sustainable*, p. 203.

13 Ibid.

14 Ibid., p. 231.

15 'Stiglitz Speaks at Occupy Wall Street – Bwog', *Bwog – Columbia Student
News* (blog), 3 October 2011, https://bwog.com/2011/10/stiglitz-spe
aks-at-occupy-wall-street/.

Chapter 5

1 'Lost in Translation (2003) – IMDb', accessed 26 October 2023, https://
www.imdb.com/title/tt0335266/.

2 Zenaida Willison, 'Civil Society Address to the WCDR', https://www.uni
sdr.org/2005/wcdr/intergover/civil-society/NGO-CDR.pdf.

3 UNISDR, 'Hyogo Framework for Action Summary' (UNISDR, 2005),
https://www.preventionweb.net/files/8720_summaryHFP20052015.pdf.

4 Willison, 'Civil Society Address to the WCDR'.

5 U.S. Mission Geneva, 'Explanation of Position of the United States for the
Sendai Framework for Disaster Risk Reduction 2015–2030', U.S. Mission
to International Organizations in Geneva, 19 March 2015, https://geneva.
usmission.gov/2015/03/19/sendai-framework-for-disaster-risk-reduct
ion-2015-2030/.

6 UNISDR, 'Sendai Framework Summary Chart' (UNISDR), accessed 26
October 2023, https://www.preventionweb.net/files/44983_sendaiframewo
rkchart.pdf.

7 Ibid.

8 Lucy Pearson and Mark Pelling, 'The UN Sendai Framework for Disaster
Risk Reduction 2015–2030: Negotiation Process and Prospects for Science
and Practice', *Journal of Extreme Events* 2, no. 1 (August 2015): 1571001,
https://doi.org/10.1142/S2345737615710013.

9 Ibid.

10 United Nations, 'The Paris Agreement', United Nations (United Nations),
accessed 7 December 2023, https://www.un.org/en/climatechange/paris-
agreement.

11 'Will the World Bank Make Good on the Loss and Damage Fund?',
ODI: Think change, 8 December 2023, https://odi.org/en/insights/will-th
e-world-bank-make-good-on-the-loss-and-damage-fund/.

12 United Nations, 'Transforming Our World: The 2030 Agenda for
Sustainable Development', 2015, https://documents-dds-ny.un.org/doc/
UNDOC/GEN/N15/291/89/PDF/N1529189.pdf?OpenElement.

Chapter 6

1 Ministere de la Culture, France, 'The Chauvet-Pont d'Arc Cave', accessed
27 October 2023, https://archeologie.culture.gouv.fr/chauvet/en. *And*
personal visit 2023.

2 Statista. 2024. 'Global Population from Ten Thousand BC to Twenty Fifty'.
https://www.statista.com/statistics/1006502/global-population-ten-thous
and-bc-to-2050/.

3 UN-DESA, 'World Population Prospects – Population Division – United
Nations', 2022, https://population.un.org/wpp/.

4 Will Steffen, Paul J. Crutzen and John R. McNeill, 'The Anthropocene: Are
Humans Now Overwhelming the Great Forces of Nature', *AMBIO: A
Journal of the Human Environment* 36, no. 8 (December 2007): 614–21,
https://doi.org/10.1579/0044-7447(2007)36[614:TAAHNO]2.0.CO;2.

5 Brad de Long, *Slouching Towards Utopia* (New York: Basic Books,
2023), https://basicbooks.uk/titles/brad-de-long/slouching-towards-uto
pia/9781399803434/.

6 Beck, *Risk Society.*

7 Christopher Gerrard, Paolo Forlin, and Peter Brown, *Waiting for the End
of the World?: New Perspectives on Natural Disasters in Medieval Europe*,
2020, https://doi.org/10.4324/9781003023449.

8 Ibid.

9 Bas van Bavel, Daniel Curtis, Jessica Dijkman, Matthew Hannaford, Maïka
De Keyzer, Eline Van Onacker and Tim Soens, *Disasters and History: The
Vulnerability and Resilience of Past Societies* (Cambridge: Cambridge
University Press, 2020), https://doi.org/10.1017/9781108569743.

10 Katherine Worboys, 'The Uses of History in Disaster Preparedness: The
1755 Lisbon Earthquake and the Construction of Historical Memory',
Journal of the International Institute 13, no. 2 (Winter 2006), http://hdl.han
dle.net/2027/spo.4750978.0013.203.

11 Stephen Broadberry et al., 'British Economic Growth, 1270–1870', 1
January 2011.

12 Ministere de la Culture, France., 'The Chauvet-Pont d'Arc Cave'.

13 Steffen et al., 'The Anthropocene'.

14 Ibid.

15 T. Toivanen, K. Lummaa, A. Majava, P. Järvensivu, V. Lähde, T. Vaden and
J. T. Eronen, 'The Many Anthropocenes: A Transdisciplinary Challenge
for the Anthropocene Research', *The Anthropocene Review* 4, no. 3 (1
December 2017): 183–98, https://doi.org/10.1177/2053019617738

099; Will Steffen, 'Introducing the Anthropocene: The Human Epoch', *Ambio* 50, no. 10 (October 2021): 1784–87, https://doi.org/10.1007/s13 280-020-01489-4

16 Tim Jackson, *Material Concerns: Pollution, Profit and Quality of Life* (Abingdon: Routledge, 1996).

17 Vittorio Valli, 'The Three Waves of the Fordist Model of Growth and the Case of China', SSRN Scholarly Paper (Rochester, NY, 1 June 2009), https://doi.org/10.2139/ssrn.1486113.

18 Steffen, 'Introducing the Anthropocene'.

19 Brazil, Russia, India, China and South Africa.

20 Nicholas Crafts and Kevin Hjortshøj O'Rourke, 'Twentieth Century Growth', *Handbook of Economic Growth* 2 (2014): 263–346, https://doi.org/10.1016/B978-0-444-53538-2.00006-X.

21 G. Dale, 'The Great Acceleration: Is It Ending and What Comes Next?', 30 June 2023, http://bura.brunel.ac.uk/handle/2438/26772.

Chapter 7

1 Field visit. Shenzhen, China, 2003.

2 'Socialist Workers', *The Economist*, accessed 28 October 2023, https://www.economist.com/finance-and-economics/2010/06/10/socialist-workers.

3 Jackson, *Material Concerns*, p. 36.

4 David Fowler, Peter Brimblecombe, John Burrows, Mathew R.Heal, Peringe Grennfelt, David S.Stevenson, Alan Jowett, Eiko Nemitz, Mhairi Coyle, Xuejun Liu, Yunhua Chang, Gary W.Fuller, Mark A.Sutton, Zbigniew Klimont, Mike H.Unsworth and Massimo Vieno, 'A Chronology of Global Air Quality', *Philosophical Transactions of the Royal Society A: Mathematical, Physical and Engineering Sciences* 378, no. 2183 (28 September 2020): 20190314, https://doi.org/10.1098/rsta.2019.0314.

5 Gregory Clark, 'The British Industrial Revolution, 1760–1860', accessed 28 October 2023, https://faculty.econ.ucdavis.edu/faculty/gclark/ecn110b/readings/ecn110b-chapter2-2005.pdf.

6 Stephen Mosley, *The Chimney of the World: A History of Smoke Pollution in Victorian and Edwardian Manchester* (Abingdon: Routledge, 2008), https://www.routledge.com/The-Chimney-of-the-World-A-History-of-Smoke-Pollution-in-Victorian-and/Mosley/p/book/9780415477673.

7 Lowell J. Satre, 'After the Match Girls' Strike: Bryant and May in the 1890s', *Victorian Studies* 26, no. 1 (1982): 7–31.

8 Andrew Meiklejohn, 'History of Lung Diseases of Coal Miners in Great Britain: Part II, 1875–1920', *British Journal of Industrial Medicine* 9, no. 2 (1952): 93–8.

9 M. Harper, 'Occupational Health Aspects of the Arsenic Extractive Industry in Britain (1868–1925)', *British Journal of Industrial Medicine*

  45, no. 9 (September 1988): 602–5, https://doi.org/10.1136/oem.45.9.602.
  Pers. Comm. 2022. National Trust, Botallack, Cornwall.

10 Jackson, *Material Concerns*.

11 Ibid.

12 Keith Laybourn, *A History of British Trade Unionism*
  (Stroud: Sutton, 1997).

13 Robert C Allen, *The Industrial Revolution: A Very Short Introduction (Very Short Introductions)* (Oxford: Oxford University Press, 2019).

14 'Later Factory Legislation', accessed 28 October 2023, https://www.parliament.uk/about/living-heritage/transformingsociety/livinglearning/19thcentury/overview/laterfactoryleg/.

15 Arkwright Society, 'Child Workers at Cromford Mill Study Pack', accessed 28 October 2023, https://www.cromfordmills.org.uk/wp-content/uploads/2021/11/Source-2-Child-Workers.pdf.

16 Allen, *The Industrial Revolution*.

17 Arthur McIvor, 'Guardians of Workers' Bodies? Trade Unions and the History of Occupational Health and Safety', *Labour History* 119, no. 1 (November 2020): 1–30, https://doi.org/10.3828/jlh.2020.16.

18 Smithsonian Magazine and Erin Blakemore, 'New Evidence Shows Peppered Moths Changed Color in Sync With the Industrial Revolution', *Smithsonian Magazine*, accessed 28 October 2023, https://www.smithsonianmag.com/smart-news/new-evidence-peppered-moths-changed-color-sync-industrial-revolution-180959282/.

19 Mosley, *The Chimney of the World*.

20 Ibid.

21 W. Walker Hanlon, 'Pollution and Mortality in the 19th Century' (Cambridge, MA: National Bureau of Economic Research, October 2015), https://doi.org/10.3386/w21647.

22 Fowler et al., 'A Chronology of Global Air Quality'.

23 Mosley, *The Chimney of the World*.

24 Brian J. Day, 'The Moral Intuition of Ruskin's "Storm-Cloud"', *Studies in English Literature, 1500–1900* 45, no. 4 (2005): 917–33.

25 UNISDR, *Making Development Sustainable*.

Chapter 8

 1 Valli, 'The Three Waves of the Fordist Model of Growth and the Case of China'.

 2 G. Frederick Thompson, 'Fordism, Post-Fordism, and the Flexible System of Production', accessed 29 October 2023, https://www.cddc.vt.edu/digitalfordism/fordism_materials/thompson.htm.

 3 Frank Knight, *The Economic Organisation* (Chicago: University of Chicago Press, 1933).

4   Paul Samuelson, *Economics: An Introductory Analysis* (New York: McGraw Hill, 1948).

5   'Advertising the Model T – The Henry Ford Blog – Blog – The Henry Ford', accessed 12 November 2023, https://www.thehenryford.org/explore/blog/advertising-the-model-t.

6   'Advertisement for the 1924 Ford Model T, "Freedom for the Woman Who Owns a Ford" – The Henry Ford', accessed 10 November 2023, https://www.thehenryford.org/artifact/32320/.

7   'American Advertising: A Brief History', accessed 29 October 2023, https://historymatters.gmu.edu/mse/ads/amadv.html.

8   U.S. Department of Labor, '100 Years of U.S. Consumer Spending: Data for the Nation, New York City, and Boston', 2006, https://www.google.co.uk/books/edition/100_Years_of_U_S_Consumer_Spending/2JnaJj1Ej KoC?hl=en&gbpv=1&printsec=frontcover.

9   Eric Rauchway, 'An Economic History of the United States 1900–1950', in *A Companion to the Modern American Novel 1900–1950*, edi John T. Matthews (Hoboken: John Wiley, 2009), 1–12, https://doi.org/10.1002/9781444310726.ch1.

10 Dave Wheelock, 'The Great Depression: An Overview' (The Federal Reserve Bank of St Louis, 2007); Gene Smiley, *Rethinking the Great Depression* (Ivan R Dee, 2003), https://rowman.com/ISBN/9781566634 717/Rethinking-the-Great-Depression; Rauchway, 'An Economic History of the United States 1900–1950'.

11 William Lockeretz, 'The Lessons of the Dust Bowl: Several Decades before the Current Concern with Environmental Problems, Dust Storms Ravaged the Great Plains, and the Threat of More Dust Storms Still Hangs over Us', *American Scientist* 66, no. 5 (1978): 560–69.

12 'Mass Exodus From the Plains | American Experience | PBS', accessed 29 October 2023, https://www.pbs.org/wgbh/americanexperience/features/surviving-the-dust-bowl-mass-exodus-plains/.

13 Rauchway, 'An Economic History of the United States 1900–1950'.

14 Ibid.

15 Brian J. McCammack, 'The American City and Environmental Pollution', in *Oxford Research Encyclopedia of American History*, 2018, https://doi.org/10.1093/acrefore/9780199329175.013.597.

16 Arnold Reitze, 'The Legislative History of US Air Pollution Control', *Houston Law Review* 36, no. 3 (1999): 679–741.

17 Ibid.

18 D. Rosner and G. Markowitz, 'A "Gift of God"?: The Public Health Controversy over Leaded Gasoline during the 1920s', *American Journal of Public Health* 75, no. 4 (April 1985): 344–52, https://doi.org/10.2105/AJPH.75.4.344.

19 Ibid.

20 Paul Krugman, *The Return of Depression Economics and the Crisis of 2008* (London: Allen Lane, 2008).

21 Richard Hornbeck, 'The Enduring Impact of the American Dust Bowl: Short- and Long-Run Adjustments to Environmental Catastrophe', *American Economic Review* 102, no. 4 (June 2012): 1477–1507, https://doi.org/10.1257/aer.102.4.1477.

22 Tom Rosentiel, 'How a Different America Responded to the Great Depression', *Pew Research Center* (blog), 14 December 2010, https://www.pewresearch.org/2010/12/14/how-a-different-america-responded-to-the-great-depression/.

Chapter 9

1   'Kathmandu, Nepal Metro Area Population 1950–2023', accessed 7 November 2023, https://www.macrotrends.net/cities/21928/kathmandu/population.

2   'Fifth Avenue – Apple Store', Apple, accessed 31 October 2023, https://www.apple.com/retail/fifthavenue/.

3   'Apple Gangnam Will Welcome First Customers This Friday, March 31 in South Korea', Apple Newsroom, accessed 31 October 2023, https://www.apple.com/newsroom/2023/03/apple-gangnam-will-welcome-first-customers-this-friday-march-31-in-south-korea/.

4   'Apple Central World Opens Friday in Thailand', Apple Newsroom (United Kingdom), accessed 31 October 2023, https://www.apple.com/uk/newsroom/2020/07/apple-central-world-opens-friday-in-thailand/.

5   'The New Apple Sanlitun Opens Today', Apple Newsroom, accessed 31 October 2023, https://www.apple.com/newsroom/2020/07/the-new-apple-sanlitun-opens-today/.

6   'World GDP 1960–2023', accessed 30 October 2023, https://www.macrotrends.net/countries/WLD/world/gdp-gross-domestic-product.

7   'World Bank Open Data Global Manufacturing', World Bank Open Data, accessed 30 October 2023, https://data.worldbank.org. https://data.worldbank.org/indicator/NV.IND.MANF.CD?end=2022&start=1997&view=chart.

8   'World Bank Open Data Poverty Head Count', World Bank Open Data, accessed 30 October 2023, https://data.worldbank.org. https://data.worldbank.org/topic/11

9   Saarah Ghazi, 'Tale of Five Cities – Risers and Fallers in the Urban League Table', *Oxford Economics*, 16 February 2023, https://www.oxfordeconomics.com/resource/tale-of-five-cities-risers-and-fallers-in-the-urban-league-table/.

10 'Cairo, Egypt Metro Area Population 1950–2023', accessed 31 October 2023, https://www.macrotrends.net/cities/22812/cairo/population.

11 Julia Bello-Schu, 'African Urban Futures', *African Futures*, no. 20 (2016), https://ethz.ch/content/dam/ethz/special-interest/gess/cis/center-for-sec urities-studies/resources/docs/ISS%20Africa%20af20.pdf.

12 'Nairobi Wealthier than 29 Counties Combined', Business Daily, 9 October 2023, https://www.businessdailyafrica.com/bd/economy/nairobi-wealth ier-than-29-counties-combined--4395478.

13 'Kenya – Gross Domestic Product (GDP) 2028', Statista, accessed 10 November 2023, https://www.statista.com/statistics/451111/gross-domes tic-product-gdp-in-kenya/.

14 Eyder Peralta, 'From Nairobi, A Rare, Clear Glimpse Of Mount Kenya Drives Disbelief On Social Media', *NPR*, 14 April 2020, sec. The Coronavirus Crisis, https://www.npr.org/sections/coronavirus-live- updates/2020/04/14/833996289/from-nairobi-a-rare-clear-glim pse-of-mt-kenya-drives-disbelief-on-social-media.

15 Charis Goodyear, 'The Conservationist Who Started a Twitter Trend and Saved Nairobi's Wildlife', University of Cambridge, 22 April 2021, https:// www.cam.ac.uk/thiscambridgelife/theconservationistwhosavednairobisw ildlife.

16 Ajit Singh, William R. Avis and Francis D. Pope, 'Visibility as a Proxy for Air Quality in East Africa', *Environmental Research Letters* 15, no. 8 (July 2020): 084002, https://doi.org/10.1088/1748-9326/ab8b12.

17 Priyanka deSouza, 'Air Pollution in Kenya: A Review', *Air Quality, Atmosphere & Health* 13, no. 12 (December 2020): 1487–95, https://doi. org/10.1007/s11869-020-00902-x.

18 'Nairobi-City-Climate-Action-2021.Pdf', accessed 11 November 2023, https://cdn.nation.co.ke/downloads/Nairobi-City-Climate-Act ion-2021.pdf.

19 'Nairobi Expressway: One Year On', accessed 10 November 2023, //global. chinadaily.com.cn/a/202308/04/WS64ccbaf7a31035260b81a5d7.html.

20 'How Nairobi's "Road for the Rich" Resulted in Thousands of Homes Reduced to Rubble', *The Guardian*, 8 December 2021, sec. Global development, https://www.theguardian.com/global-development/2021/ dec/08/how-nairobis-road-for-the-rich-resulted-in-thousands-of-homes- reduced-to-rubble.

21 Eliya M. Zulu, Donatien Beguy, Alex C. Ezeh, Philippe Bocquier, Nyovani J. Madise, John Cleland and Jane Falkingham, 'Overview of Migration, Poverty and Health Dynamics in Nairobi City's Slum Settlements', *Journal of Urban Health: Bulletin of the New York Academy of Medicine* 88, no. Suppl 2 (June 2011): 185–99, https://doi.org/10.1007/s11524-011-9595-0.

22 'World Bank Country and Lending Groups – World Bank Data Help Desk', accessed 3 November 2023, https://datahelpdesk.worldbank.org/ knowledgebase/articles/906519.

23 Maarten van Ham, Tiit Tammaru, Rūta Ubarevičienė and Heleen Janssen, 'Rising Inequalities and a Changing Social Geography of Cities. An

Introduction to the Global Segregation Book', in *Urban Socio-Economic Segregation and Income Inequality: A Global Perspective*, ed. Maarten van Ham, Tiit Tammaru, Rūta Ubarevičienė and Heleen Janssen, The Urban Book Series (Cham: Springer International, 2021), 3–26, https://doi.org/10.1007/978-3-030-64569-4_1.

24 Julia Bird, Piero Montebruno and Tanner Regan, 'Life in a Slum: Understanding Living Conditions in Nairobi's Slums across Time and Space', *Oxford Review of Economic Policy* 33, no. 3 (2017): 496–520, https://doi.org/10.1093/oxrep/grx036.

25 'Some Facts and Stats about Kibera, Kenya | Kibera UK', 21 June 2015, https://www.kibera.org.uk/facts-info/; Bird, Montebruno, and Regan, 'Life in a Slum'.

26 'The_case_of_kibera_edited.Pdf', accessed 1 November 2023, https://unhabitat.org/sites/default/files/2021/08/the_case_of_kibera_edited.pdf.

27 Field visit March 2008.

28 'SDG Indicators', accessed 30 October 2023, https://unstats.un.org/sdgs/report/2019/goal-11/.

29 Bird et al., 'Life in a Slum'.

30 'Africa Droughts Became More Frequent, More Intense and Widespread over the Last Four Decades – WaterAid | WaterAid UK', accessed 10 November 2023, https://www.wateraid.org/uk/media/africa-droughts-became-more-frequent-more-intense-and-widespread-over-the-last-four-decades.

31 Bangkok Post Public Company Limited, 'Designing Your Own Cooking Space', *Bangkok Post*, accessed 10 November 2023, https://www.bangkokpost.com/life/social-and-lifestyle/1434674/designing-your-own-cooking-space.

32 'City Mayors: Richest Cities in the World in 2020 by GDP', accessed 3 November 2023, http://www.citymayors.com/statistics/richest-cities-2020.html.

33 'Subnational HDI – Subnational HDI – Table – Global Data Lab', accessed 10 November 2023, https://globaldatalab.org/shdi/shdi/THA/?levels=1%2B4&interpolation=1&extrapolation=0&nearest_real=0&years=2019.

34 Apiwat Ratanawaraha, 'Inequality, Fragility, and Resilience in Bangkok', Building Resilience in Cities under Stress (New York: International Peace Institute, 2016), https://www.jstor.org/stable/resrep09526.5.

35 Ibid.

36 'A Decade on, Learning from Thailand's Devastating 2011 Floods | Swiss Re', accessed 3 November 2023, https://www.swissre.com/risk-knowledge/mitigating-climate-risk/decade-on-thailand-devastating-2011-floods.html.

37 'Thai Floods Batter Global Electronics, Auto Supply Chains', *Reuters*, 28 October 2011, sec. Environment, https://www.reuters.com/article/us-thai-floods-idUSTRE79R0QR20111028.

38 Ibid.; 'A Decade on, Learning from Thailand's Devastating 2011 Floods | Swiss Re'.

39 Ibid.

40 Rebecca Ratcliffe and Rebecca Ratcliffe South-East Asia correspondent, 'Bangkok Air Pollution Prompts Advice to Work from Home', *The Guardian*, 26 January 2023, sec. World news, https://www.theguardian.com/world/2023/jan/26/bangkok-air-pollution-prompts-advice-to-work-from-home.

41 'Bangkok Smog Blankets City Hospitalising 200,000', euronews, 10 March 2023, https://www.euronews.com/green/2023/03/10/its-very-hard-for-me-to-breathe-lately-thick-smog-covers-bangkok-and-hospitalises-thousand.

42 Nuntavarn Vichit-Vadakan and Nitaya Vajanapoom, 'Health Impact from Air Pollution in Thailand: Current and Future Challenges', *Environmental Health Perspectives* 119, no. 5 (May 2011): A197, https://doi.org/10.1289/ehp.1103728.

43 'Air Pollution in Bangkok: Addressing Unequal Exposure and Enhancing Public Understanding of the Risks', 7 March 2023, https://doi.org/10.51414/sei2023.12.

44 Atipon Satranarakun and Tanpat Kraiwanit, 'Factors Affecting Travel in the Bangkok Metropolitan Region', *Asian Journal of Applied Economics* 29, no. 2 (7 November 2022): 71–91.

45 U.N. Environment, 'Cities and Climate Change', UNEP – UN Environment Programme, 26 September 2017, http://www.unep.org/explore-topics/resource-efficiency/what-we-do/cities/cities-and-climate-change.

46 'Big New Road Projects in Bangkok Suburbs – Rama 2 Special Road Set for Completion in 2025', Thailand News, Travel & Forum – ASEAN NOW, 3 February 2022, https://aseannow.com/topic/1248959-big-new-road-projects-in-bangkok-suburbs-rama-2-special-road-set-for-completion-in-2025/.

47 'Thailand: Total Revenue of Bangkok 2021', Statista, accessed 3 November 2023, https://www.statista.com/statistics/1063524/thailand-total-revenue-of-bangkok/.

48 James Crabtree, 'Le Corbusier's Chandigarh: An Indian City Unlike Any Other', *Financial Times*, 3 July 2015, sec. Financial Times, https://www.ft.com/content/2a194cb4-1a8d-11e5-a130-2e7db721f996.

49 'Chandigarh', in *Wikipedia*, 2 November 2023, https://en.wikipedia.org/w/index.php?title=Chandigarh&oldid=1183120508.

50 'Subnational HDI – Subnational HDI – Table – Global Data Lab'.

51 'Chandigarh', 2 November 2023.

52   Peter Fitting, 'Urban Planning/Utopian Dreaming: Le Corbusier's
     Chandigarh Today', *Utopian Studies* 13, no. 1 (2002): 69–93.
53   Khaiwal Ravindra, Tanbir Singh, Vivek Pandey and Suman Mor,
     'Air Pollution Trend in Chandigarh City Situated in Indo-Gangetic
     Plains: Understanding Seasonality and Impact of Mitigation Strategies',
     Science of The Total Environment 729 (10 August 2020): 138717, https://
     doi.org/10.1016/j.scitotenv.2020.138717.
54   'Slumisation of Chandigarh: A Bugbear That's Hard to Tame', *Hindustan
     Times*, 13 May 2022, https://www.hindustantimes.com/cities/chandig
     arh-news/slumisation-of-chandigarh-a-bugbear-that-s-hard-to-tame-1016
     52386943092.html.
55   Manoj Kumar Teotia, 'Housing for the Urban Poor in Chandigarh:
     Including the Excluded', accessed 10 November 2023, https://www.acade
     mia.edu/16915001/Housing_for_the_Urban_Poor_in_Chandigarh_Inc
     luding_the_Excluded; Namita Gupta and Kavita, 'Slum Rehabilitation
     Through Public Housing Schemes in India: A Case of Chandigarh',
     Environment and Urbanization ASIA 11, no. 2 (1 September 2020): 231–
     46, https://doi.org/10.1177/0975425320938536
56   Ravindra et al., 'Air Pollution Trend in Chandigarh City Situated in Indo-
     Gangetic Plains'; 'City – Chandigarh, India', accessed 10 November 2023,
     https://urbanemissions.info/india-apna/chandigarh-india/.
57   'Mohali's Mega Makeover! How These Key Infra Projects Will Ease
     Traffic Congestion across City', *Financialexpress* (blog), 27 March 2023,
     https://www.financialexpress.com/business/roadways-mohalis-mega-
     makeover-how-these-key-infra-projects-will-ease-traffic-congestion-acr
     oss-city-3023227/.
58   'Just 25 Mega-Cities Produce 52% of the World's Urban Greenhouse Gas
     Emissions – Science & Research News | Frontiers', 12 July 2021, https://
     blog.frontiersin.org/2021/07/12/just-25-mega-cities-produce-52-of-the-
     worlds-urban-greenhouse-gas-emissions/.
59   Department for Transport, 'Latest Evidence on Induced Travel
     Demand: An Evidence Review', 2018, https://assets.publishing.service.gov.
     uk/media/5c0e5848e5274a0bf3cbe124/latest-evidence-on-induced-travel-
     demand-an-evidence-review.pdf.
60   James K. Mitchell, 'Megacities and Natural Disasters: A Comparative
     Analysis*', *GeoJournal* 49, no. 2 (1 October 1999): 137–42, https://
     doi.org/10.1023/A:1007024703844; David Satterthwaite and Sheridan
     Bartlett, 'Editorial: The Full Spectrum of Risk in Urban Centres: Changing
     Perceptions, Changing Priorities', *Environment and Urbanization* 29, no. 1
     (1 April 2017): 3–14, https://doi.org/10.1177/0956247817691921; L Bull-
     Kamanga et al., 'From Everyday Hazards to Disasters: The Accumulation
     of Risk in Urban Areas', *Environment and Urbanization* 15, no. 1 (1 April
     2003): 193–204, https://doi.org/10.1177/095624780301500109.

61 Cairo (Egypt), Cape Town (SAR), Johannesburg (SAR), Hong Kong (Hong Kong), Jakarta (Indonesia), Mumbai (India), Shanghai (China), Tel Aviv (Israel), Tokyo (Japan), Melbourne (Australia), Berlin (Germany), Brussels (Belgium), Istanbul (Turkey), London (UK), Paris (France), Chicago (United States), Los Angeles (United States), Mexico City (Mexico), New York (United States), Bogotá (Colombia), Buenos Aires (Argentina), Lima (Peru), Paramaribo (Suriname) and São Paulo (Brazil).

62 Long, *Slouching Towards Utopia*.

63 Ibid.

64 Mitchell, 'Megacities and Natural Disasters'; Satterthwaite and Bartlett, 'Editorial'; Bull-Kamanga et al., 'From Everyday Hazards to Disasters'; Ben Wisner, 'Disaster Risk Reduction in Megacities: Making the Most of Human and Social Capital', *Building Safer Cities: The Future of Disaster Risk, Disaster Risk Management Series* 13 (1 January 2003).

65 'World Rural Population 1960–2023', accessed 2 November 2023, https://www.macrotrends.net/countries/WLD/world/rural-population.

66 Sabina Alkire, Mihika Chatterje, Adriana Conconi, Suman Seth and Ana Vaz, 'Poverty in Rural and Urban Areas: Direct Comparisons Using the Global MPI 2014' (University of Oxford, June 2014), https://doi.org/10.35648/20.500.12413/11781/ii020.

67 Venla Niva, Alexander Horton, Vili Virkki, Matias Heino, Maria Kosonen, Marko Kallio, Pekka Kinnunen, Guy J. Abel, Raya Muttarak, Maija Taka, Olli Varis and Matti Kummu, 'World's Human Migration Patterns in 2000–2019 Unveiled by High-Resolution Data', *Nature Human Behaviour*, 7 September 2023, 1–15, https://doi.org/10.1038/s41562-023-01689-4.

68 'To Move the Needle on Ending Extreme Poverty, Focus on Rural Areas', Brookings, accessed 2 November 2023, https://www.brookings.edu/articles/to-move-the-needle-on-ending-extreme-poverty-focus-on-rural-areas/.

69 'The Green Revolution: Accomplishments and Apprehensions', accessed 8 November 2023, https://www.agbioworld.org/biotech-info/topics/borlaug/borlaug-green.html.

70 Hannah Ritchie, Pablo Rosado and Max Roser, 'Environmental Impacts of Food Production', *Our World in Data*, 2 December 2022, https://ourworldindata.org/environmental-impacts-of-food.

71 'Global Agribusiness Continues to Displace Rural Communities', 3 July 2021, https://www.landportal.org/node/100007.

72 Andrés Cadena, Richard Dobbs and Jaana Remes, 'The Growing Economic Power of Cities', *Journal of International Affairs* 65, no. 2 (2012): 1–17.

73 'World GDP 1960–2023'.

74 Environment, 'Cities and Climate Change'.

75 'Global CO_2 Emissions by Year 1940–2022'.

76 Steffen et al., 'The Anthropocene'.

77 Martin J. Head, Will Steffen, David Fagerlind, Colin N. Waters, Clement Poirier, Jaia Syvitski, Jan A. Zalasiewicz et al., 'The Great Acceleration Is Real and Provides a Quantitative Basis for the Proposed Anthropocene Series/Epoch', *Episodes* 45, no. 4 (2021): 359.

78 Katherine Calvin, Dipak Dasgupta, Gerhard Krinner, Aditi Mukherji, Peter W. Thorne, Christopher Trisos, José Romero et al., 'IPCC, 2023: Climate Change 2023: Synthesis Report. Contribution of Working Groups I, II and III to the Sixth Assessment Report of the Intergovernmental Panel on Climate Change [Core Writing Team, H. Lee and J. Romero (Eds.)]. IPCC, Geneva, Switzerland.', First (Intergovernmental Panel on Climate Change (IPCC), 25 July 2023), https://doi.org/10.59327/IPCC/AR6-9789291691647.

79 Herman E. Daly, 'Toward Some Operational Principles of Sustainable Development', *Ecological Economics* 2, no. 1 (1 April 1990): 1–6, https://doi.org/10.1016/0921-8009(90)90010-R.

80 Johan Rockström, Will Steffen, Kevin Noone, Åsa Persson, F. Stuart Iii Chapin, Eric Lambin, Timothy M. Lenton et al., 'Planetary Boundaries: Exploring the Safe Operating Space for Humanity', *Ecology and Society* 14, no. 2 (2009): art32, https://doi.org/10.5751/ES-03180-140232.

81 Robert Monroe, 'The History of the Keeling Curve', The Keeling Curve, 3 April 2013, https://keelingcurve.ucsd.edu/2013/04/03/the-history-of-the-keeling-curve/; Monroe, 'The History of the Keeling Curve'; Monroe, 'The History of the Keeling Curve'.

Chapter 10

1 'Main Findings and Recommendations of the Midterm Review of the Implementation of the Sendai Framework for Disaster Risk Reduction 2015–2030', accessed 1 December 2023, https://documents-dds-ny.un.org/doc/UNDOC/GEN/N22/764/26/PDF/N2276426.pdf?OpenElement.

2 Ibid.

3 United Nations Department of Economic and Social Affairs, *The Sustainable Development Goals Report 2023: Special Edition*, The Sustainable Development Goals Report (United Nations, 2023), https://doi.org/10.18356/9789210024914.

4 U.N. Environment, 'Emissions Gap Report 2023', UNEP – UN Environment Programme, 11 August 2023, http://www.unep.org/resources/emissions-gap-report-2023.

5 'Outcome of the First Global Stocktake. Draft Decision -/CMA.5. Proposal by the President | UNFCCC', accessed 13 December 2023, https://unfccc.int/documents/636608.

6 Nina Lakhani and Nina Lakhani Climate justice reporter, 'Record Number of Fossil Fuel Lobbyists Get Access to Cop28 Climate Talks', *The Guardian*, 5 December 2023, sec. Environment, https://www.theguardian.com/envi ronment/2023/dec/05/record-number-of-fossil-fuel-lobbyists-get-acc ess-to-cop28-climate-talks.

7 'Top 50 Global Reinsurance Groups – Reinsurance News', ReinsuranceNe. ws, accessed 15 December 2023, https://www.reinsurancene.ws/top-50-reinsurance-groups/.

8 'The Economics of Climate Change'.

9 Ibid.

10 Ibid.

11 John Burn-Murdoch, 'What We Get Wrong When We Talk about Global Warming', *Financial Times*, 21 July 2023, sec. Data Points, https://www. ft.com/content/de449d0d-0558-48da-90b9-5bb4fe809dab.

12 Ibid.

13 Ilan Kelman, *Disaster by Choice: How Our Actions Turn Natural Hazards into Catastrophes* (Oxford: Oxford University Press, 2020).

Chapter 11

1 Atlanta Symphony Orchestra, *Durufle and Faure: Requiems*, CD (Telarc, 1987), https://www.prestomusic.com/classical/products/7951205--duru fle-faure-requiems.

2 UNISDR, *Making Development Sustainable*.

3 UN-Habitat, 'The New Urban Agenda', Habitat III, 2016, https://habitat3. org/the-new-urban-agenda/.

4 'Main Findings and Recommendations of the Midterm Review of the Implementation of the Sendai Framework for Disaster Risk Reduction 2015–2030'.

5 Kate Raworth, *Doughnut Economics* (London: Penguin, 2017).

6 Daly, 'Toward Some Operational Principles of Sustainable Development'.

7 Rockström et al., 'Planetary Boundaries'.

8 Anna Zalik and Isaac Asume' Osuoka, 'Beyond Transparency: A Consideration of Extraction's Full Costs', *The Extractive Industries and Society* 7, no. 3 (July 2020): 781–5, https://doi.org/10.1016/j. exis.2020.07.015.

9 Jackson, *Material Concerns*.

10 'Corporate Scandals That Changed the Course of Capitalism | Baillie Gifford', accessed 5 January 2024, https://www.bailliegifford.com/en/uk/ individual-investors/insights/ic-video/2019-q4-corporate-scandals-that-changed-the-course-of-capitalism-all-we-0295/; 'What Was the Rana Plaza Disaster and Why Did It Happen?', *The Independent*, 23 April 2020,

https://www.independent.co.uk/life-style/fashion/rana-plaza-factory-disaster-anniversary-what-happened-fashion-a9478126.html.

11 Nicolai J. Foss, *Theories of the Firm: Critical Perspectives in Economic Organisation (3 Vols.). Introductory Chapter* (London, Routledge, 1999).

12 Hendry, 'An Introduction to Theories of the Firm'.

13 Brian Roach, *Corporate Power in a Global Economy*, Economics in Context Initiative, Global Development Policy Center (Boston: Boston University, 2023), https://www.bu.edu/eci/files/2023/09/Corporate-Power-Module.pdf.

14 Nicholas Stern, Joseph Stiglitz and Charlotte Taylor, 'The Economics of Immense Risk, Urgent Action and Radical Change: Towards New Approaches to the Economics of Climate Change', *Journal of Economic Methodology* 29, no. 3 (3 July 2022): 181–216, https://doi.org/10.1080/1350178X.2022.2040740.

15 'Sustainable Products At B&Q – Key Milestones | Buying Standards | Responsible Business | B&Q', accessed 21 December 2023, https://www.diy.com/responsible-business/products/buying-standards/sustainable-products.

16 Field visits, Ecuador, India, the Philippines and Hong Kong, 1998–2002.

17 Milton Friedman, 'A Friedman Doctrine – The Social Responsibility of Business Is to Increase Its Profits', *The New York Times*, 13 September 1970, sec. Archives, https://www.nytimes.com/1970/09/13/archives/a-friedman-doctrine-the-social-responsibility-of-business-is-to.html.

18 Guido Orzes, Antonella Moretto, Mattia Moro, Matteo Rossi, Marco Sartor, FedericoCaniato and Guido Nassimbeni, 'The Impact of the United Nations Global Compact on Firm Performance: A Longitudinal Analysis', *International Journal of Production Economics* 227 (1 February 2020): 107664, https://doi.org/10.1016/j.ijpe.2020.107664.

19 Robert Armstrong, 'The Fallacy of ESG Investing', *Financial Times*, 23 October 2020, sec. Advice & Comment, https://www.ft.com/content/9e3e1d8b-bf9f-4d8c-baee-0b25c3113319.

20 Charl de Villiers, Matteo La Torre and Matteo Molinari, 'The Global Reporting Initiative's (GRI) Past, Present and Future: Critical Reflections and a Research Agenda on Sustainability Reporting (Standard-Setting)', SSRN Scholarly Paper (Rochester, NY, 2022), https://papers.ssrn.com/abstract=4099915.

21 Anjli Raval, 'The Struggle for the Soul of the B Corp Movement', *Financial Times*, 19 February 2023, sec. The Big Read, https://www.ft.com/content/0b632709-afda-4bdc-a6f3-bb0b02eb5a62.

22 Larry Elliott, 'IMF Warns That Tech Giants Stifle Innovation and Threaten Stability', *The Guardian*, 3 April 2019, sec. Business, https://www.theguardian.com/business/2019/apr/03/imf-warns-that-tech-giants-stifle-innovation-and-threaten-stability.

23 Alan Greene, *Emergency Powers in a Time of Pandemic* (Bristol: Bristol University Press, 2020).

24 'A World At Risk' (Global Preparedness Monitoring Board, 2019), https://www.gpmb.org/docs/librariesprovider17/default-document-library/annual-reports/gpmb-2019-annualreport-en.pdf?sfvrsn=d1c9143c_30.

25 UNDRR, 'Financing Prevention and De-Risking Investment', 2021, https://www.undrr.org/sites/default/files/2022-12/P2136%20Policy%20Brief_Financing_prevention_and_de-risking_investment_DIGITAL.pdf.

26 Milton Friedman, '"Free Trade"', *Newsweek*, 17 August 1970.

27 UK Government, 'Red Tape Challenge Update: Implementation Plan for the Retail Theme', 2013, https://assets.publishing.service.gov.uk/media/5a796e0940f0b63d72fc5be9/13-560-red-tape-challenge-update-implementation-plan-for-retail-theme.pdf; 'One-in, Two-out: Statement of New Regulation', GOV.UK, 30 December 2014, https://www.gov.uk/government/collections/one-in-two-out-statement-of-new-regulation.

28 'Government's "One in, Two Out" Deregulation Rule Suspended for Grenfell Response', Inside Housing, accessed 2 January 2024, https://www.insidehousing.co.uk/news/governments-one-in-two-out-deregulation-rule-suspended-for-grenfell-response-60217.

29 Robert Booth and Robert Booth Social affairs correspondent, 'Grenfell Inquiry Told Government Had Ideological Aversion to Red Tape', *The Guardian*, 30 March 2022, sec. UK news, https://www.theguardian.com/uk-news/2022/mar/30/grenfell-inquiry-told-government-had-ideological-aversion-to-red-tape.

30 'The "Forever War" on Red Tape and the Struggle to Improve Regulation', accessed 26 October 2023, https://onlinelibrary.wiley.com/doi/10.1111/1467-8500.12534.

31 Steven K. Vogel, 'Rethinking Stigler's Theory of Regulation: Regulatory Capture or Deregulatory Capture?', *ProMarket* (blog), 15 May 2018, https://www.promarket.org/2018/05/15/rethinking-stiglers-theory-regulation-regulatory-capture-deregulatory-capture/.

32 Joseph Persky and Herbert Tsang, 'Pigouvian Exploitation of Labor', *The Review of Economics and Statistics* 56, no. 1 (1974): 52–7, https://doi.org/10.2307/1927526.

33 Joseph, 'Natural Capital at Risk: The Top 100 Externalities of Business', Capitals Coalition, 1 April 2013, https://capitalscoalition.org/natural-capital-at-risk-the-top-100-externalities-of-business/.

34 Sam Peltzman, Michael E. Levine and Roger G. Noll, 'The Economic Theory of Regulation after a Decade of Deregulation', *Brookings Papers on Economic Activity. Microeconomics* 1989 (1989): 1–59, https://doi.org/10.2307/2534719.

35 Inge Kaul, Isabelle Grunberg and Marc A. Stern, eds, *Global Public Goods: International Cooperation in the 21st Century* (New York: Oxford University Press, 1999).

36 'Global Governance and Governance of the Global Commons in the Global Partnership for Development beyond 2015', accessed 11 March 2024, https://www.un.org/en/development/desa/policy/untaskteam_undf/thinkpieces/24_thinkpiece_global_governance.pdf.

37 Jinseop Jang, Jason McSparren and Yuliya Rashchupkina, 'Global Governance: Present and Future', *Palgrave Communications* 2, no. 1 (19 January 2016): 1–5, https://doi.org/10.1057/palcomms.2015.45.

38 Peter Nadin, 'The United Nations: A History of Success and Failure', *AQ: Australian Quarterly* 90, no. 4 (2019): 11–17.

39 Raworth, *Doughnut Economics*.

40 Pew Research Center, 'Emerging and Developing Economies Much More Optimistic Than Rich Countries about the Future', *Pew Research Center's Global Attitudes Project* (blog), 9 October 2014, https://www.pewresearch.org/global/2014/10/09/emerging-and-developing-economies-much-more-optimistic-than-rich-countries-about-the-future/.

41 Bill Gates, *How to Avoid a Climate Disaster* (London: Penguin, 2022).

42 'A Short History of Enclosure in Britain | The Land Magazine', accessed 19 December 2023, https://www.thelandmagazine.org.uk/articles/short-hist ory-enclosure-britain.

43 'John Locke: The Justification of Private Property | Libertarianism.Org', 19 October 2015, https://www.libertarianism.org/columns/john-locke-justif ication-private-property.

44 Fred P. Saunders, 'The Promise of Common Pool Resource Theory and the Reality of Commons Projects', 8, no. 2 (31 August 2014): 636, https://doi.org/10.18352/ijc.477.

45 Joseph, 'Natural Capital at Risk'.

46 Anna Salvi and Clara Krimm, 'The Commons: A Key Concept for a New Pact between Mankind and Nature', *Revue interdisciplinaire d'études juridiques* 85, no. 2 (2020): 243–70, https://doi.org/10.3917/riej.085.0243.

47 Elinor Ostrom, *Governing the Commons: The Evolution of Institutions for Collective Action*, Canto Classics (Cambridge: Cambridge University Press, 2015), https://doi.org/10.1017/CBO9781316423936.

48 'Green Governance: Ecological Survival, Human Rights and the Commons | The Wealth of the Commons', accessed 18 December 2023, https://wealthofthecommons.org/essay/green-governance-ecological-survi val-human-rights-and-commons.

49 Sheila R. Foster and Christian Iaione, 'The City as a Commons', *Yale Law & Policy Review* 34, no. 2 (2016): 281–349; Christian Iaione, 'The Co-City: Sharing, Collaborating, Cooperating, and Commoning in the City', *The American Journal of Economics and Sociology* 75, no. 2 (2016): 415–55; Jonathan D. Rosenbloom, 'New Day at the Pool: State Preemption, Common Pool Resources, and Non-Place Based Municipal Collaborations', SSRN Scholarly Paper (Rochester, NY, 28 July 2011), https://papers.ssrn.com/abstract=1898270.

50 Richard Heeks, 'Grassroots ICT4D Innovation', *ICTs for Development* (blog), 13 June 2009, https://ict4dblog.wordpress.com/2009/06/13/grassro ots-ict4d-innovation/.

Chapter 12

1 Robert Putnam, *Bowling Alone* (New York: Simon and Schuster, 2001).
2 Krista Halttunen, Raphael Slade, and Iain Staffell, ' "We Don't Want to Be the Bad Guys": Oil Industry's Sensemaking of the Sustainability Transition Paradox', *Energy Research & Social Science* 92 (1 October 2022): 102800, https://doi.org/10.1016/j.erss.2022.102800.
3 'German Farmers Blockade Berlin with Tractors in Subsidy Row', *BBC News*, 8 January 2024, sec. Europe, https://www.bbc.com/news/world-eur ope-67911739.
4 'Life-Threatening Heat Is Forcing Germany to Ramp Up Defenses', *Bloomberg.Com*, 12 August 2023, https://www.bloomberg.com/news/artic les/2023-08-12/extreme-heat-is-forcing-germany-to-ramp-up-defenses.
5 'New Years Eve Tractor Run from Higher Farm Equine Byley – Cardiac Risk in the Young', 8 January 2024, https://www.c-r-y.org.uk/ new-years-eve-tractor-run-from-higher-farm-equine-byley/.
6 'Public Support Majority of Net Zero Policies … Unless There Is a Personal Cost'.
7 Ibid.
8 'Climate Change Concerns Make Many Around the World Willing to Alter How They Live and Work | Pew Research Center', accessed 21 November 2023, https://www.pewresearch.org/global/2021/09/14/ in-response-to-climate-change-citizens-in-advanced-economies-are-will ing-to-alter-how-they-live-and-work/.
9 Anthony Leiserowitz, Edward Maibach, Connie Roser-Renouf, Seth Rosenthal, Matthew Cutler and John Kotcher, 'Politics and Global Warming' (Yale Program on Climate Change Communication, 2018), https://climatecommunication.yale.edu/publications/politics-global-warm ing-march-2018/.
10 'Climate Change – July 2023 – Eurobarometer Survey', accessed 22 December 2023, https://europa.eu/eurobarometer/surveys/detail/2954.
11 'World's Largest Survey of Public Opinion on Climate Change: A Majority of People Call for Wide-Ranging Action', UNDP, accessed 8 January 2024, https://www.undp.org/press-releases/worlds-largest-survey-public-opin ion-climate-change-majority-people-call-wide-ranging-action.
12 Peter Andre, Teodora Boneva, Felix Chopra and Armin Falk, 'Globally Representative Evidence on the Actual and Perceived Support for Climate Action', *Nature Climate Change* 14, no. 3 (March 2024): 253–9, https://doi. org/10.1038/s41558-024-01925-3.

13 Ting Liu, Nick Shryane and Mark Elliot, 'Attitudes to Climate Change Risk: Classification of and Transitions in the UK Population between 2012 and 2020', *Humanities and Social Sciences Communications* 9, no. 1 (18 August 2022): 1–15, https://doi.org/10.1057/s41599-022-01287-1.

14 Anthony G. Patt and Elke U. Weber, 'Perceptions and Communication Strategies for the Many Uncertainties Relevant for Climate Policy', *WIREs Climate Change* 5, no. 2 (2014): 219–32, https://doi.org/10.1002/wcc.259.

15 'Governments Face Losing the Battle against Climate Change | Chatham House – International Affairs Think Tank', 15 February 2022, https://www.chathamhouse.org/2022/02/governments-face-losing-battle-against-climate-change.

16 Juliette N. Rooney-Varga, Margaret Hensel, Carolyn McCarthy, Karen McNeal, Nicole Norfles, Kenneth Rath, Audrey H. Schnell and John D. Sterman, 'Building Consensus for Ambitious Climate Action through the *World Climate* Simulation', *Earth's Future* 9, no. 12 (December 2021): 1–16, https://doi.org/10.1029/2021EF002283.

17 Esther Michelsen Kjeldahl and Vincent F Hendricks, 'The Sense of Social Influence: Pluralistic Ignorance in Climate Change: Social Factors Play Key Roles in Human Behavior. Individuals Tend to Underestimate How Much Others Worry about Climate Change. This May Inhibit Them from Taking Collective Climate Action', *EMBO Reports* 19, no. 11 (November 2018): e47185, https://doi.org/10.15252/embr.201847185.

18 Sanna Inthorn, Justin Lewis and Karin Wahl-Jorgensen, 'Images of Citizenship on Television News', *Journalism Studies* 5, no. 2 (2004): 153–64.

19 Jon Alexander, *Citizens* (Kingston on Thames: Canbury Press, 2023), https://www.canburypress.com/products/citizens-by-jon-alexander.

20 Paulo Freire, *Pedagogy of the Oppressed*, Revised edition 1996 (London: Penguin, 1970).

21 Robert Chambers, *Rural Development: Putting the Last First* (Abingdon: Routledge, 1983), https://www.routledge.com/Rural-Development-Putting-the-last-first/Chambers/p/book/9780582644434.

22 Gates, *How to Avoid a Climate Disaster*.

23 Ann Pettifor, *The Case for the Green New Deal* (London: Verso, 2019).

24 Raworth, *Doughnut Economics*.

25 Sustainable Development Commission, 'Prosperity without Growth? – The Transition to a Sustainable Economy Sustainable Development Commission', 2009, https://www.sd-commission.org.uk/publications.php@id=914.html.

26 Jason Hickel, Giorgos Kallis, Tim Jackson, Daniel W. O'Neill, Juliet B. Schor, Julia K. Steinberger, Peter A. Victor and Diana Ürge-Vorsatz, 'Degrowth Can Work – Here's How Science Can Help', *Nature* 612, no. 7940 (December 2022): 400–3, https://doi.org/10.1038/d41586-022-04412-x.

27 'Beating the Abatement Cost Curve for Growth | McKinsey', accessed 12 January 2024, https://www.mckinsey.com/capabilities/operations/our-insights/net-zero-or-bust-beating-the-abatement-cost-curve-for-growth.

28 Ben Wisner, 'Five Years Beyond Sendai – Can We Get Beyond Frameworks?', *International Journal of Disaster Risk Science* 11, no. 2 (1 April 2020): 239–49, https://doi.org/10.1007/s13753-020-00263-0.

29 Christiana Figueres and Tom Rivett-Carnac, *The Future We Choose* (London: Manilla Press, 2020).

30 Ibid.

31 Kenneth Teitelbaum, Myles Horton, Paulo Friere, Brenda Bell, John Gaventa and John Peters, 'We Make the Road by Walking: Conversations on Education and Social Change', *History of Education Quarterly* 32, no. 1 (1992): 146, https://doi.org/10.2307/368424.

32 UN-Habitat, 'The New Urban Agenda'.

33 Simon Elias Bibri, John Krogstie and Mattias Kärrholm, 'Compact City Planning and Development: Emerging Practices and Strategies for Achieving the Goals of Sustainability', *Developments in the Built Environment*, 2020, https://doi.org/10.1016/j.dibe.2020.100021; 'The Interconnected City: Imagining Our Urban Lives in 2050', Metabolic, accessed 8 January 2024, https://www.metabolic.nl/news/the-interconnec ted-city/.

34 Kangning Huang, Xia Li, Xiaoping Liu and Karen C. Seto, 'Projecting Global Urban Land Expansion and Heat Island Intensification through 2050', *Environmental Research Letters* 14, no. 11 (1 November 2019): 114037, https://doi.org/10.1088/1748-9326/ab4b71.

35 Bibri, Krogstie and Kärrholm, 'Compact City Planning and Development'.

36 'The Interconnected City'.

37 Jack Strauss, Hongchang Li and Jinli Cui, 'High-Speed Rail's Impact on Airline Demand and Air Carbon Emissions in China', *Transport Policy* 109 (1 August 2021): 85–97, https://doi.org/10.1016/j.tranpol.2021.05.019.

38 Heinrich Bofinger and Jon Strand, *Calculating the Carbon Footprint from Different Classes of Air Travel*, Policy Research Working Papers (The World Bank, 2013), https://doi.org/10.1596/1813-9450-6471.

39 'FACT FOCUS: Conspiracies Misconstrue "15-Minute City" Idea', AP News, 3 March 2023, https://apnews.com/article/fact-check-15-min ute-city-conspiracy-162fd388f0c435a8289cc9ea213f92ee.

40 Esther Addley, '"This Is Political Expediency": How the Tories Turned on 15-Minute Cities', *The Guardian*, 7 October 2023, sec. Cities, https://www. theguardian.com/cities/2023/oct/07/15-minute-cities-rishi-sunak-tories-conspiracy-theory.

41 Bibri et al., 'Compact City Planning and Development'.

42 'Eyes on the Street: East L.A. Median Stormwater Capture Project, Multiuse Path – Streetsblog Los Angeles', 11 June 2021, https://la.streetsb

log.org/2021/06/11/eyes-on-the-street-east-la-median-stormwater-capt
ure-project-multiuse-path.

43 'How Valencia Turned A Crisis (And a River) into a Transformative Park', *Metropolis* (blog), accessed 1 April 2024, https://metropolismag.com/proje cts/how-valencia-turned-crisis-river-into-park/.

Chapter 13

1 'Views from Frontline Report 2011 | Resources at GNDR', *GNDR* (blog), accessed 12 January 2024, https://www.gndr.org/resource/views-from-the-frontline/views-from-the-frontline-report-2011/.

2 David Mosse, 'Is Good Policy Unimplementable? Reflections on the Ethnography of Aid Policy and Practice', *Development and Change* 35, no. 4 (2004): 639–71, https://doi.org/10.1111/j.0012-155X.2004.00374.x.

3 Steve Waddell, *Global Action Networks* (London: Palgrave Macmillan UK, 2011).

4 Brendan Cox, 'Campaigning for International Justice' (Bond, 2011), https:// inventing-futures.org/wp-content/uploads/2024/01/Campaigning_for_ International_Justice_Brendan_Cox_May_2011.pdf.

5 @NatGeoUK, '26 Hopeful Visions for a Sustainable Future', National Geographic, 4 November 2021, https://www.nationalgeographic. co.uk/26visions.

Conclusion: Unnatural disasters, convenient untruths

1 Weiwei, 'Our Duty Is to Remember Sichuan'.

2 *An Inconvenient Truth*, DVD (Paramount, 2006).

3 UNDRR, 'The Report of the Midterm Review of the Implementation of the Sendai Framework for Disaster Risk Reduction 2015–2030 | Midterm Review of the Sendai Framework', 5 April 2023, http://sendaiframew ork-mtr.undrr.org/publication/report-midterm-review-implementation-sendai-framework-disaster-risk-reduction-2015-2030.

4 United Nations Department of Economic and Social Affairs, *The Sustainable Development Goals Report 2023*.

SELECT BIBLIOGRAPHY

Alexander, Jon. *Citizens*. Kingston on Thames: Canbury Press, 2023. https://www.canburypress.com/products/citizens-by-jon-alexander.

Al-Hajj, Samar, Hassan R. Dhaini, Stefania Mondello, Haytham Kaafarani, Firas Kobeissy and Ralph G. DePalma. 'Beirut Ammonium Nitrate Blast: Analysis, Review, and Recommendations'. *Frontiers in Public Health* 9 (2021). https://www.frontiersin.org/articles/10.3389/fpubh.2021.657996.

Alkire, Sabina, Mihika Chatterje, Adriana Conconi, Suman Seth and Ana Vaz. 'Poverty in Rural and Urban Areas: Direct Comparisons Using the Global MPI 2014'. University of Oxford, June 2014. https://doi.org/10.35648/20.500.12413/11781/ii020.

Allen, Robert C. *The Industrial Revolution: A Very Short Introduction (Very Short Introductions)*. Oxford: Oxford University Press, 2019.

Andre, Peter, Teodora Boneva, Felix Chopra and Armin Falk. 'Globally Representative Evidence on the Actual and Perceived Support for Climate Action'. *Nature Climate Change* 14, no. 3 (March 2024): 253–9. https://doi.org/10.1038/s41558-024-01925-3.

Bari, Sarwar. 'Waiting for Politics at the Mercy of River: Case Study of an Enduring Community'. *Disaster Prevention and Management: An International Journal* 28, no. 1 (1 January 2018): 42–9. https://doi.org/10.1108/DPM-06-2018-0187.

Bavel, Bas, Daniel Curtis, Jessica Dijkman, Matthew Hannaford, Maïka De Keyzer, Eline Van Onacker and Tim Soens. *Disasters and History: The Vulnerability and Resilience of Past Societies*. Cambridge: Cambridge University Press, 2020. https://doi.org/10.1017/9781108569743.

Beck, Ulrich. *Risk Society*. London: SAGE Publications, 1992.

Bello-Schu, Julia. 'African Urban Futures'. *African Futures*, no. 20 (2016). https://ethz.ch/content/dam/ethz/special-interest/gess/cis/center-for-sec urities-studies/resources/docs/ISS%20Africa%20af20.pdf.

Bibri, Simon Elias, John Krogstie and Mattias Kärrholm. 'Compact City Planning and Development: Emerging Practices and Strategies for Achieving the Goals of Sustainability'. *Developments in the Built Environment* 4 (2020): 100021. https://doi.org/10.1016/j.dibe.2020.100021.

Bird, Julia, Piero Montebruno and Tanner Regan. 'Life in a Slum: Understanding Living Conditions in Nairobi's Slums across Time and Space'. *Oxford Review of Economic Policy* 33, no. 3 (2017): 496–520. https://doi.org/10.1093/oxrep/grx036.

Blaikie, Piers, Terry Cannon, Ian Davis and Ben Wisner. *At Risk: Natural Hazards, People's Vulnerability and Disasters*. 2nd edn. London: Routledge, 2014. https://doi.org/10.4324/9780203714775.

Bofinger, Heinrich, and Jon Strand. *Calculating the Carbon Footprint from Different Classes of Air Travel*. Policy Research Working Papers. The World Bank, 2013. https://doi.org/10.1596/1813-9450-6471.

Broadberry, Stephen, Bruce Campbell, Alexander Klein, Mark Overton and Bas Leeuwen. 'British Economic Growth, 1270–1870: An Output-Based Approach', 1 January 2011.

Bull-Kamanga, L., K. Diagne, A. Lavell, E. Leon, F. Lerise, H. MacGregor, A. Maskrey et al. 'From Everyday Hazards to Disasters: The Accumulation of Risk in Urban Areas'. *Environment and Urbanization* 15, no. 1 (1 April 2003): 193–204. https://doi.org/10.1177/095624780301500109.

Burton, Ian. 'Forensic Disaster Investigations in Depth: A New Case Study Model'. *Environment: Science and Policy for Sustainable Development* 52, no. 5 (31 August 2010): 36–41. https://doi.org/10.1080/00139157.2010.507144.

Cadena, Andrés, Richard Dobbs and Jaana Remes. 'The Growing Economic Power of Cities'. *Journal of International Affairs* 65, no. 2 (2012): 1–17.

Calvin, Katherine, Dipak Dasgupta, Gerhard Krinner, Aditi Mukherji, Peter W. Thorne, Christopher Trisos, José Romero et al. 'IPCC, 2023: Climate Change 2023: Synthesis Report. Contribution of Working Groups I, II and III to the Sixth Assessment Report of the Intergovernmental Panel on Climate Change [Core Writing Team, H. Lee and J. Romero (Eds.)]. IPCC, Geneva, Switzerland.' First. Intergovernmental Panel on Climate Change (IPCC), 25 July 2023. https://doi.org/10.59327/IPCC/AR6-9789291691647.

Chambers, Robert. *Rural Development: Putting the Last First*. Abingdon: Routledge, 1983. https://www.routledge.com/Rural-Development-Putting-the-last-first/Chambers/p/book/9780582644434.

Cox, Brendan. 'Campaigning for International Justice'. Bond, 2011. https://inventing-futures.org/wp-content/uploads/2024/01/Campaigning_for_International_Justice_Brendan_Cox_May_2011.pdf.

Crafts, Nicholas, and Kevin Hjortshøj O'Rourke. 'Twentieth Century Growth'. *Handbook of Economic Growth* 2 (2014): 263–346. https://doi.org/10.1016/B978-0-444-53538-2.00006-X.

Cuny, Frederick C. *Disaster and Development Cuny*. New York: Oxford University Press, 1983.

Dale, G. 'The Great Acceleration: Is It Ending and What Comes Next?', 30 June 2023. http://bura.brunel.ac.uk/handle/2438/26772.

Daly, Herman E. 'Toward Some Operational Principles of Sustainable Development'. *Ecological Economics* 2, no. 1 (1 April 1990): 1–6. https://doi.org/10.1016/0921-8009(90)90010-R.

Davis, Ian. *Shelter After Disaster*. Oxford: Oxford Polytechnic Press, 1978. http://shelterprojects.org/other/siteplanning/site-planning-references/Davis-1978-shelter%20after%20disaster.pdf.

Day, Brian J. 'The Moral Intuition of Ruskin's "Storm-Cloud"'. *Studies in English Literature, 1500–1900* 45, no. 4 (2005): 917–33.

De Vivo, Benedetto, and Giuseppe Rolandi. 'Vesuvius: Volcanic Hazard and Civil Defense'. *Rendiconti Lincei* 24, no. 1 (March 2013): 39–45. https://doi.org/10.1007/s12210-012-0212-2.

deSouza, Priyanka. 'Air Pollution in Kenya: A Review'. *Air Quality, Atmosphere & Health* 13, no. 12 (December 2020): 1487–95. https://doi.org/10.1007/s11869-020-00902-x.

Figueres, Christiana, and Tom Rivett-Carnac. *The Future We Choose*. London: Manilla Press, 2020.

Fitting, Peter. 'Urban Planning/Utopian Dreaming: Le Corbusier's Chandigarh Today'. *Utopian Studies* 13, no. 1 (2002): 69–93.

Foss, Nicolai J. *Theories of the Firm: Critical Perspectives in Economic Organisation (3 Vols.). Introductory Chapter.* London: Routledge, 1999.

Foster, Sheila R., and Christian Iaione. 'The City as a Commons'. *Yale Law & Policy Review* 34, no. 2 (2016): 281–349.

Fowler, David, Peter Brimblecombe, John Burrows, Mathew R. Heal, Peringe Grennfelt, David S. Stevenson, Alan Jowett, et al. 'A Chronology of Global Air Quality'. *Philosophical Transactions of the Royal Society A: Mathematical, Physical and Engineering Sciences* 378, no. 2183 (28 September 2020): 20190314. https://doi.org/10.1098/rsta.2019.0314.

Freire, Paulo. *Pedagogy of the Oppressed*. Revised edition 1996. London: Penguin, 1970.

Frieberg, A. 'The "Forever War" on Red Tape and the Struggle to Improve Regulation'. Accessed 26 October 2023. https://onlinelibrary.wiley.com/doi/10.1111/1467-8500.12534.

Frühling, Pierre. 'Turning Disasters into Opportunities'. SIDA, May 2002. https://cdn.sida.se/publications/files/sida2152en-turning-disasters-into-opportunities.pdf.

Gallin, Luke. 'Lower Cats, but Elevated Attritional Losses to Hit P&C Re/Insurers in Q1: Analysts – Reinsurance News'. ReinsuranceNe.ws, 8 April 2019. http://www.reinsurancene.ws/lower-cats-but-elevated-attritional-losses-to-hit-pc-re-insurers-in-q1-analysts/.

Gates, Bill. *How to Avoid a Climate Disaster*. London: Penguin, 2022.

Gerrard, Christopher, Paolo Forlin and Peter Brown. *Waiting for the End of the World? New Perspectives on Natural Disasters in Medieval Europe*, 2020. https://doi.org/10.4324/9781003023449.

Gibson, Terry. 'Frontline: From Local Information to Local Resilience'. GNDR, 2017. https://www.gndr.org/wp-content/uploads/2021/11/FRONTLINE_PUBLICATION-English.pdf.

Gibson, Terry. 'Learning How to Build Back Better'. *Monday Developments: Interaction USA*, June 2010.

Gibson, Terry. *Making Aid Agencies Work: Reconnecting INGOs with the People They Serve*. Leeds: Emerald Publishing Limited, 2019. https://doi.org/10.1108/9781787695092.

Gostin, Lawrence O., and Gigi K. Gronvall. 'The Origins of Covid-19 – Why It Matters (and Why It Doesn't)'. *New England Journal of Medicine* 388, no. 25 (22 June 2023): 2305–8. https://doi.org/10.1056/NEJMp2305081.

Greene, Alan. *Emergency Powers in a Time of Pandemic*. Bristol: Bristol University Press, 2020.

Griffith, David. 'Migration, Labor Scarcity, and Deforestation in Honduran Cattle Country'. *Journal of Ecological Anthropology* 18, no. 1 (December 2016). https://doi.org/10.5038/2162-4593.18.1.3.

Gupta, Namita, and Kavita. 'Slum Rehabilitation Through Public Housing Schemes in India: A Case of Chandigarh'. *Environment and Urbanization ASIA* 11, no. 2 (1 September 2020): 231–46. https://doi.org/10.1177/09754 25320938536.

Halttunen, Krista, Raphael Slade and Iain Staffell. '"We Don't Want to Be the Bad Guys": Oil Industry's Sensemaking of the Sustainability Transition Paradox'. *Energy Research & Social Science* 92 (1 October 2022): 102800. https://doi.org/10.1016/j.erss.2022.102800.

Ham, Maarten, Tiit Tammaru, Rūta Ubarevičienė and Heleen Janssen. 'Rising Inequalities and a Changing Social Geography of Cities. An Introduction to the Global Segregation Book'. In *Urban Socio-Economic Segregation and Income Inequality: A Global Perspective*, edited by Maarten Ham, Tiit Tammaru, Rūta Ubarevičienė and Heleen Janssen, 3–26. The Urban Book Series. Cham: Springer International, 2021. https://doi.org/10.1007/978-3-030-64569-4_1.

Hanlon, W. Walker. 'Pollution and Mortality in the 19th Century'. Cambridge, MA: National Bureau of Economic Research, October 2015. https://doi.org/10.3386/w21647.

Harper, M. 'Occupational Health Aspects of the Arsenic Extractive Industry in Britain (1868–1925)'. *British Journal of Industrial Medicine* 45, no. 9 (September 1988): 602–5. https://doi.org/10.1136/oem.45.9.602.

Hazarika, H., N. P. Bhandary, Y. Kajita, K. Kasama, K. Tsukahara and R. K. Pokharel. 'The 2015 Nepal Gorkha Earthquake: An Overview of the Damage, Lessons Learned and Challenges'. *Lowland Technology International* 18, no. 2 (2016): 105–18.

Head, Martin J., Will Steffen, David Fagerlind, Colin N. Waters, Clement Poirier, Jaia Syvitski, Jan A. Zalasiewicz, et al. 'The Great Acceleration Is Real and Provides a Quantitative Basis for the Proposed Anthropocene Series/Epoch'. *Episodes* 45, no. 4 (2021): 359.

Hickel, Jason, Giorgos Kallis, Tim Jackson, Daniel W. O'Neill, Juliet B. Schor, Julia K. Steinberger, Peter A. Victor and Diana Ürge-Vorsatz. 'Degrowth Can Work – Here's How Science Can Help'. *Nature* 612, no. 7940 (December 2022): 400–3. https://doi.org/10.1038/d41586-022-04412-x.

Hornbeck, Richard. 'The Enduring Impact of the American Dust Bowl: Short-
and Long-Run Adjustments to Environmental Catastrophe'. *American
Economic Review* 102, no. 4 (June 2012): 1477–1507. https://doi.
org/10.1257/aer.102.4.1477.

House, Paul. 'Forest Farmers: A Case Study of Traditional Shifting Cultivation
in Honduras'. *Network Paper-Rural Development Forestry Network* (…, 1
January 1997. https://www.academia.edu/666608/Forest_Farmers_A_case_
study_of_traditional_shifting_cultivation_in_Honduras.

Huang, Kangning, Xia Li, Xiaoping Liu and Karen C. Seto. 'Projecting Global
Urban Land Expansion and Heat Island Intensification through 2050'.
Environmental Research Letters 14, no. 11 (1 November 2019): 114037.
https://doi.org/10.1088/1748-9326/ab4b71.

Hübl, Thomas. *Healing Collective Trauma: A Process for Integrating Our
Intergenerational & Cultural Wounds*. Boulder, CO: Sounds True, 2020.

Iaione, Christian. 'The Co-City: Sharing, Collaborating, Cooperating, and
Commoning in the City'. *The American Journal of Economics and Sociology*
75, no. 2 (2016): 415–55.

Inthorn, Sanna, Justin Lewis and Karin Wahl-jorgensen. 'Images of Citzenship
on Television News'. *Journalism Studies* 5, no. 2 (2004): 153–64.

Ishigaki, Akemi, Hikari Higashi, Takako Sakamoto and Shigeki Shibahara.
'The Great East-Japan Earthquake and Devastating Tsunami: An Update
and Lessons from the Past Great Earthquakes in Japan since 1923'. *The
Tohoku Journal of Experimental Medicine* 229, no. 4 (April 2013): 287–99.
https://doi.org/10.1620/tjem.229.287.

Jackson, Tim. *Material Concerns: Pollution, Profit and Quality of Life*.
Abingdon: Routledge, 1996.

Jang, Jinseop, Jason McSparren and Yuliya Rashchupkina. 'Global
Governance: Present and Future'. *Palgrave Communications* 2, no. 1 (19
January 2016): 1–5. https://doi.org/10.1057/palcomms.2015.45.

Jennings, Steve. 'Time's Bitter Flood: Trends in the Number of Reported
Natural Disasters'. OXFAM, 1 May 2011. https://www.researchgate.net/publ
ication/313847052_Time's_Bitter_Flood_Trends_in_the_number_of_rep
orted_natural_disasters.

Kaul, Inge, Isabelle Grunberg and Marc A. Stern, eds. *Global Public Goods:
International Cooperation in the 21st Century*. New York: Oxford University
Press, 1999.

Kelman, Ilan. *Disaster by Choice: How Our Actions Turn Natural Hazards into
Catastrophes*. Oxford: Oxford University Press, 2020.

Kerle, Norman, Benjamin Van Wyk De Vries and Clive Oppenheimer. 'New
Insight into the Factors Leading to the 1998 Flank Collapse and Lahar
Disaster at Casita Volcano, Nicaragua'. *Bulletin of Volcanology* 65, no. 5
(July 2003): 331–45. https://doi.org/10.1007/s00445-002-0263-9.

Keys, Patrick W., Victor Galaz, Michelle Dyer, Nathanial Matthews, Carl
Folke, Magnus Nyström and Sarah E. Cornell. 'Anthropocene Risk'. *Nature*

Sustainability 2, no. 8 (22 July 2019): 667–73. https://doi.org/10.1038/s41 893-019-0327-x.

Kjeldahl, Esther Michelsen, and Vincent F Hendricks. 'The Sense of Social Influence: Pluralistic Ignorance in Climate Change: Social Factors Play Key Roles in Human Behavior. Individuals Tend to Underestimate How Much Others Worry about Climate Change. This May Inhibit Them from Taking Collective Climate Action'. *EMBO Reports* 19, no. 11 (November 2018): e47185. https://doi.org/10.15252/embr.201847185.

Knight, Frank. *The Economic Organisation*. Chicago: University of Chicago Press, 1933.

Krugman, Paul. *The Return of Depression Economics and the Crisis of 2008*. London: Allen Lane, 2008.

Laybourn, Keith. *A History of British Trade Unionism*. Stroud: Sutton, 1997.

Leiserowitz, A., E. Maibach, C. Roser-Renouf, S. Rosenthal, M. Cutler and J. Kotcher. 'Politics and Global Warming'. Yale Program on Climate Change Communication, 2018. https://climatecommunication.yale.edu/publicati ons/politics-global-warming-march-2018/.

Liu, Ting, Nick Shryane and Mark Elliot. 'Attitudes to Climate Change Risk: Classification of and Transitions in the UK Population between 2012 and 2020'. *Humanities and Social Sciences Communications* 9, no. 1 (18 August 2022): 1–15. https://doi.org/10.1057/s41599-022-01287-1.

Lockeretz, William. 'The Lessons of the Dust Bowl: Several Decades before the Current Concern with Environmental Problems, Dust Storms Ravaged the Great Plains, and the Threat of More Dust Storms Still Hangs over Us'. *American Scientist* 66, no. 5 (1978): 560–9.

Long, Brad de. *Slouching Towards Utopia*. New York: Basic Books, 2023. https://basicbooks.uk/titles/brad-de-long/slouching-towards-uto pia/9781399803434/.

Martin, Stephanie C. 'Past Eruptions and Future Predictions: Analyzing Ancient Responses to Mount Vesuvius for Use in Modern Risk Management'. *Journal of Volcanology and Geothermal Research* 396 (1 May 2020): 106851. https://doi.org/10.1016/j.jvolgeores.2020.106851.

McCammack, Brian J. 'The American City and Environmental Pollution'. In *Oxford Research Encyclopedia of American History*, 2018. https://doi. org/10.1093/acrefore/9780199329175.013.597.

McIvor, Arthur. 'Guardians of Workers' Bodies? Trade Unions and the History of Occupational Health and Safety'. *Labour History* 119, no. 1 (November 2020): 1–30. https://doi.org/10.3828/jlh.2020.16.

McSweeney, Kendra, and Oliver T. Coomes. 'Climate-Related Disaster Opens a Window of Opportunity for Rural Poor in Northeastern Honduras'. *Proceedings of the National Academy of Sciences* 108, no. 13 (29 March 2011): 5203–8. https://doi.org/10.1073/pnas.1014123108.

Meiklejohn, Andrew. 'History of Lung Diseases of Coal Miners in Great Britain: Part II, 1875–1920'. *British Journal of Industrial Medicine* 9, no. 2 (1952): 93–8.

Merrill, Tim. *Honduras: A Country Study*. Library of Congress, 1993. https://babel.hathitrust.org/cgi/pt?id=mdp.39015031871166&view=1up&seq=159.

Mitchell, James K. 'Megacities and Natural Disasters: A Comparative Analysis*'. *GeoJournal* 49, no. 2 (1 October 1999): 137–42. https://doi.org/10.1023/A:1007024703844.

Miyamoto, H. Kit, and Tom Chan. 'Assessment of Damage and Lessons Learned'. *Structure Magazine*, January 2009. https://www.structuremag.org/wp-content/uploads/2014/08/C-Str-Perf_China_Earthquake-Miyamoto-Gilani-Chan-Jan-09.pdf.

Mosley Stephen. *The Chimney of the World: A History of Smoke Pollution in Victorian and Edwardian Manchester*. Abingdon: Routledge, 2008. https://www.routledge.com/The-Chimney-of-the-World-A-History-of-Smoke-Pollution-in-Victorian-and/Mosley/p/book/9780415477673.

Mosse, David. 'Is Good Policy Unimplementable? Reflections on the Ethnography of Aid Policy and Practice'. *Development and Change* 35, no. 4 (2004): 639–71. https://doi.org/10.1111/j.0012-155X.2004.00374.x.

Mukunda, Gautam. 'The Social and Political Costs of the Financial Crisis, 10 Years Later'. *Harvard Business Review*, 25 September 2018. https://hbr.org/2018/09/the-social-and-political-costs-of-the-financial-crisis-10-years-later.

Nadin, Peter. 'The United Nations: A History of Success and Failure'. *AQ: Australian Quarterly* 90, no. 4 (2019): 11–17.

Nguyen, Thi Phuoc Lai, E. Winjikul and S. G. P. Virdis. 'Air Pollution in Bangkok: Addressing Unequal Exposure and Enhancing Public Understanding of the Risks', 7 March 2023. https://doi.org/10.51414/sei2023.12.

Norton, John, and Terry David Gibson. 'Introduction to Disaster Prevention: Doing It Differently by Rethinking the Nature of Knowledge and Learning'. *Disaster Prevention and Management: An International Journal* 28, no. 1 (1 January 2019): 2–5. https://doi.org/10.1108/DPM-02-2019-323.

O'Keefe, Phil, Ken Westgate and Ben Wisner. 'Taking the Naturalness out of Natural Disasters'. *Nature* 260 (15 April 1976): 566–67. https://doi.org/10.1038/260566a0.

Oliver-Smith, A., Irasema Alcántara-Ayala, Ian Burton and Allan Lavell. *Forensic Investigations of Disasters (FORIN): A Conceptual Framework and Guide to Research*. IRDR, 2016. https://www.researchgate.net/publication/291349173_Forensic_Investigations_of_Disasters_FORIN_a_conceptual_framework_and_guide_to_research.

Oliver-Smith, Anthony, Irasema Alcántara-Ayala, Ian Burton and Allan Lavell. 'The Social Construction of Disaster Risk: Seeking Root Causes'.

International Journal of Disaster Risk Reduction 22 (June 2017): 469–74. https://doi.org/10.1016/j.ijdrr.2016.10.006.

Orzes, Guido, Antonella Moretto, Mattia Moro, Matteo Rossi, Marco Sartor, FedericoCaniato and Guido Nassimbeni. 'The Impact of the United Nations Global Compact on Firm Performance: A Longitudinal Analysis'. *International Journal of Production Economics* 227 (1 February 2020): 107664. https://doi.org/10.1016/j.ijpe.2020.107664.

Ostrom, Elinor. *Governing the Commons: The Evolution of Institutions for Collective Action*. Canto Classics. Cambridge: Cambridge University Press, 2015. https://doi.org/10.1017/CBO9781316423936.

Oxley, M. 'Pakistan Floods: Preventing Future Catastrophic Flood Disasters'. *Preventionweb*, 2010. https://www.preventionweb.net/files/15697 _01.10.101.pdf.

Patt, Anthony G., and Elke U. Weber. 'Perceptions and Communication Strategies for the Many Uncertainties Relevant for Climate Policy'. *WIREs Climate Change* 5, no. 2 (2014): 219–32. https://doi.org/10.1002/wcc.259.

Pearson, Lucy, and Mark Pelling. 'The UN Sendai Framework for Disaster Risk Reduction 2015–2030: Negotiation Process and Prospects for Science and Practice'. *Journal of Extreme Events* 2, no. 1 (August 2015): 1571001. https://doi.org/10.1142/S2345737615710013.

Peltzman, Sam, Michael E. Levine and Roger G. Noll. 'The Economic Theory of Regulation after a Decade of Deregulation'. *Brookings Papers on Economic Activity. Microeconomics* 1989 (1989): 1–59. https://doi.org/10.2307/2534719.

Persky, Joseph, and Herbert Tsang. 'Pigouvian Exploitation of Labor'. *The Review of Economics and Statistics* 56, no. 1 (1974): 52–7. https://doi.org/10.2307/1927526.

Peters, Laura E. R., Aaron Clark-Ginsberg, Bernard McCaul, Gabriela Cáceres, Ana Luisa Nuñez, Jay Balagna, Alejandra López, Sonny S. Patel, Ronak B. Patel and Jamon Van Den Hoek. 'Informality, Violence, and Disaster Risks: Coproducing Inclusive Early Warning and Response Systems in Urban Informal Settlements in Honduras'. *Frontiers in Climate* 4 (2022). https://www.frontiersin.org/articles/10.3389/fclim.2022.937244.

Pettifor, Ann. *The Case for the Green New Deal*. London: Verso, 2019.

Piracha, Awais, and Muhammad Tariq Chaudhary. 'Urban Air Pollution, Urban Heat Island and Human Health: A Review of the Literature'. *Sustainability* 14, no. 15 (January 2022): 9234. https://doi.org/10.3390/su1 4159234.

Putnam, Robert. *Bowling Alone*. New York: Simon and Schuster, 2001.

Quarantelli, E. L. 'Disaster Studies: An Analysis of the Social Historical Factors Affecting the Development of Research in the Area'. *International Journal of Mass Emergencies & Disasters* 5, no. 3 (1 November 1987): 285–310. https://doi.org/10.1177/028072708700500306.

Ratanawaraha, Apiwat. 'Inequality, Fragility, and Resilience in Bangkok'. Building Resilience in Cities under Stress. International Peace Institute, 2016. New York. https://www.jstor.org/stable/resrep09526.5.

Rauchway, Eric. 'An Economic History of the United States 1900–1950'. In John T. Matthews (ed.), *A Companion to the Modern American Novel 1900–1950*, 1–12. Hoboken: John Wiley & Sons, Ltd, 2009. https://doi.org/10.1002/9781444310726.ch1.

Ravindra, Khaiwal, Tanbir Singh, Vivek Pandey and Suman Mor. 'Air Pollution Trend in Chandigarh City Situated in Indo-Gangetic Plains: Understanding Seasonality and Impact of Mitigation Strategies'. *Science of the Total Environment* 729 (10 August 2020): 138717. https://doi.org/10.1016/j.scitotenv.2020.138717.

Raworth, Kate. *Doughnut Economics*. London: Penguin, 2017.

Reitze, Arnold. 'The Legislative History of US Air Pollution Control'. *Houston Law Review* 36, no. 3 (1999): 679–741.

Ritchie, Hannah, Pablo Rosado, and Max Roser. 'Environmental Impacts of Food Production'. *Our World in Data*, 2 December 2022. https://ourworldindata.org/environmental-impacts-of-food.

Roach, Brian. *Corporate Power in a Global Economy*. Economics in Context Initiative, Global Development Policy Center. Boston: Boston University, 2023. https://www.bu.edu/eci/files/2023/09/Corporate-Power-Module.pdf.

Rockström, Johan, Will Steffen, Kevin Noone, Åsa Persson, F. Stuart Iii Chapin, Eric Lambin, Timothy M. Lenton, et al. 'Planetary Boundaries: Exploring the Safe Operating Space for Humanity'. *Ecology and Society* 14, no. 2 (2009): art32. https://doi.org/10.5751/ES-03180-140232.

Rogers, Malcolm. 'What Happened to Honduras after Hurricane Mitch'. Christian Aid, 1999. http://web.archive.org/web/20070606090147/http://www.christianaid.org.uk/indepth/9910inde/indebt2.htm.

Rooney-Varga, Juliette N., Margaret Hensel, Carolyn McCarthy, Karen McNeal, Nicole Norfles, Kenneth Rath, Audrey H. Schnell and John D. Sterman. 'Building Consensus for Ambitious Climate Action Through the *World Climate* Simulation'. *Earth's Future* 9, no. 12 (December 2021): 1–16. https://doi.org/10.1029/2021EF002283.

Rosenbloom, Jonathan D. 'New Day at the Pool: State Preemption, Common Pool Resources, and Non-Place Based Municipal Collaborations'. SSRN Scholarly Paper. Rochester, NY, 28 July 2011. https://papers.ssrn.com/abstract=1898270.

Rosner, D., and G. Markowitz. 'A "Gift of God"? The Public Health Controversy over Leaded Gasoline during the 1920s'. *American Journal of Public Health* 75, no. 4 (April 1985): 344–52. https://doi.org/10.2105/AJPH.75.4.344.

Sabet, Daniel M. 'When Corruption Funds the Political System: A Case Study of Honduras'. Woodrow Wilson Center, 2020. https://www.wilsoncenter.

org/publication/when-corruption-funds-political-system-case-study-honduras.

Salvi, Anna, and Clara Krimm. 'The Commons: A Key Concept for a New Pact between Mankind and Nature'. *Revue interdisciplinaire d'études juridiques* 85, no. 2 (2020): 243–70. https://doi.org/10.3917/riej.085.0243.

Samuelson, Paul. *Economics: An Introductory Analysis*. New York: McGraw Hill, 1948.

Satranarakun, Atipon, and Tanpat Kraiwanit. 'Factors Affecting Travel in the Bangkok Metropolitan Region'. *Asian Journal of Applied Economics* 29, no. 2 (7 November 2022): 71–91.

Satre, Lowell J. 'After the Match Girls' Strike: Bryant and May in the 1890s'. *Victorian Studies* 26, no. 1 (1982): 7–31.

Satterthwaite, David, and Sheridan Bartlett. 'Editorial: The Full Spectrum of Risk in Urban Centres: Changing Perceptions, Changing Priorities'. *Environment and Urbanization* 29, no. 1 (1 April 2017): 3–14. https://doi.org/10.1177/0956247817691921.

Saunders, Fred P. 'The Promise of Common Pool Resource Theory and the Reality of Commons Projects', 8, no. 2 (31 August 2014): 636. https://doi.org/10.18352/ijc.477.

Scally, Gabriel, Bobbie Jacobson and Kamran Abbasi. 'The UK's Public Health Response to Covid-19'. *BMJ* 369 (15 May 2020): m1932. https://doi.org/10.1136/bmj.m1932.

Silver, Roxane Cohen, E. Alison Holman and Dana Rose Garfin. 'Coping with Cascading Collective Traumas in the United States'. *Nature Human Behaviour* 5, no. 1 (January 2021): 4–6. https://doi.org/10.1038/s41 562-020-00981-x.

Singh, Ajit, William R. Avis and Francis D. Pope. 'Visibility as a Proxy for Air Quality in East Africa'. *Environmental Research Letters* 15, no. 8 (July 2020): 084002. https://doi.org/10.1088/1748-9326/ab8b12.

Smiley, Gene. *Rethinking the Great Depression*. Ivan R. Dee, 2003. https://row man.com/ISBN/9781566634717/Rethinking-the-Great-Depression.

Smith, William C. 'Hurricane Mitch and Honduras: An Illustration of Population Vulnerability'. *International Journal of Health System and Disaster Management* 1, no. 1 (2013): 54. https://doi.org/10.4103/2347-9019.122460.

Steffen, Will. 'Introducing the Anthropocene: The Human Epoch'. *Ambio* 50, no. 10 (October 2021): 1784–87. https://doi.org/10.1007/s13 280-020-01489-4.

Steffen, Will, Paul J. Crutzen and John R. McNeill. 'The Anthropocene: Are Humans Now Overwhelming the Great Forces of Nature'. *AMBIO: A Journal of the Human Environment* 36, no. 8 (December 2007): 614–21. https://doi.org/10.1579/0044-7447(2007)36[614:TAAHNO]2.0.CO;2.

Stern, Nicholas, Joseph Stiglitz and Charlotte Taylor. 'The Economics of Immense Risk, Urgent Action and Radical Change: Towards

New Approaches to the Economics of Climate Change'. *Journal of Economic Methodology* 29, no. 3 (3 July 2022): 181–216. https://doi.org/10.1080/1350178X.2022.2040740.

Stonich, Susan C. *I Am Destroying The Land! The Political Ecology Of Poverty And Environmental Destruction In Honduras*. New York: Routledge, 2021. https://doi.org/10.4324/9780429041068.

Strauss, Jack, Hongchang Li and Jinli Cui. 'High-Speed Rail's Impact on Airline Demand and Air Carbon Emissions in China'. *Transport Policy* 109 (1 August 2021): 85–97. https://doi.org/10.1016/j.tranpol.2021.05.019.

Subramanian, Meera. 'Anthropocene Now: Influential Panel Votes to Recognize Earth's New Epoch'. *Nature*, 21 May 2019. https://doi.org/10.1038/d41586-019-01641-5.

Susman, Paul, Phil O'Keefe and Ben Wisner. 'Global Disasters: A Radical Interpretation'. *Interpretations of Calamity*, 1 January 1983, 263–83.

Sustainable Development Commission. 'Prosperity without Growth? – The Transition to a Sustainable Economy · Sustainable Development Commission', 2009. https://www.sd-commission.org.uk/publications.php@id=914.html.

Swiss Reinsurance. 'The Economics of Climate Change'. Swiss Reinsurance, April 2021. https://www.swissre.com/institute/research/topics-and-risk-dialogues/climate-and-natural-catastrophe-risk/expertise-publication-economics-of-climate-change.html.

Teitelbaum, Kenneth, Myles Horton, Paulo Friere, Brenda Bell, John Gaventa and John Peters. 'We Make the Road by Walking: Conversations on Education and Social Change'. *History of Education Quarterly* 32, no. 1 (1992): 146. https://doi.org/10.2307/368424.

Telford, John, Margaret Arnold and Alberto Harth. 'Learning Lessons from Disaster Recovery'. The World Bank, 2004. https://documents.worldbank.org/pt/publication/documents-reports/documentdetail/966021468034504310/learning-lessons-from-disaster-recovery-the-case-of-honduras.

Titz, Alexandra. 'Geographies of Doing Nothing–Internal Displacement and Practices of Post-Disaster Recovery in Urban Areas of the Kathmandu Valley, Nepal'. *Social Sciences* 10 (22 March 2021): 110. https://doi.org/10.3390/socsci10030110.

Toivanen, T., K. Lummaa, A. Majava, P. Järvensivu, V. Lähde, T. Vaden and J. T. Eronen. 'The Many Anthropocenes: A Transdisciplinary Challenge for the Anthropocene Research'. *The Anthropocene Review* 4, no. 3 (1 December 2017): 183–98. https://doi.org/10.1177/2053019617738099.

UNDRR. 'GAR2022: Our World at Risk', 2022. https://www.undrr.org/gar2022-our-world-risk-gar.

UNDRR. 'The Report of the Midterm Review of the Implementation of the Sendai Framework for Disaster Risk Reduction 2015-2030 | Midterm Review of the Sendai Framework', 5 April 2023. http://sendaiframew

ork-mtr.undrr.org/publication/report-midterm-review-implementation-sendai-framework-disaster-risk-reduction-2015-2030.

UNISDR, ed. *GAR 2015 Making Development Sustainable: The Future of Disaster Risk Management*. Global Assessment Report on Disaster Risk Reduction, 4. 2015. Geneva: United Nations, 2015. https://www.prevention web.net/english/hyogo/gar/2015/en/home/.

UNISDR. *Living with Risk: A Global Review of Disaster Reduction Initiatives*. 2004 version. New York: United Nations, 2004.

United Nations. 'Transforming Our World: The 2030 Agenda for Sustainable Development', 2015. https://documents-dds-ny.un.org/doc/UNDOC/GEN/N15/291/89/PDF/N1529189.pdf?OpenElement.

United Nations Department of Economic and Social Affairs. *The Sustainable Development Goals Report 2023: Special Edition*. The Sustainable Development Goals Report. United Nations, 2023. https://doi.org/10.18356/9789210024914.

United States Geological Survey. 'Socioeconomic and Environmental Impacts of Landslides', July 2001. https://pubs.usgs.gov/of/2001/ofr-01-0276/.

Valli, Vittorio. 'The Three Waves of the Fordist Model of Growth and the Case of China'. SSRN Scholarly Paper. Rochester, NY, 1 June 2009. https://doi.org/10.2139/ssrn.1486113.

Vichit-Vadakan, Nuntavarn, and Nitaya Vajanapoom. 'Health Impact from Air Pollution in Thailand: Current and Future Challenges'. *Environmental Health Perspectives* 119, no. 5 (May 2011): A197. https://doi.org/10.1289/ehp.1103728.

Villiers, Charl, Matteo La Torre and Matteo Molinari. 'The Global Reporting Initiative's (GRI) Past, Present and Future: Critical Reflections and a Research Agenda on Sustainability Reporting (Standard-Setting)'. SSRN Scholarly Paper. Rochester, NY, 2022. https://papers.ssrn.com/abstract=4099915.

Vogel, Steven K. 'Rethinking Stigler's Theory of Regulation: Regulatory Capture or Deregulatory Capture?' *ProMarket* (blog), 15 May 2018. https://www.promarket.org/2018/05/15/rethinking-stiglers-theory-regulation-regulatory-capture-deregulatory-capture/.

Waddell, Steve. *Global Action Networks*. London: Palgrave Macmillan UK, 2011.

Wall, Michael, and Ian Davis. 'Christian Perspectives on Disaster Management: A Training Manual'. Interchurch Relief and Development Alliance, 1992. https://desastres.unanleon.edu.ni/pdf/2002/agosto/PDF/ENG/DOC3788/DOC3788.HTM.

Wisner, Ben. 'Disaster Risk Reduction in Megacities: Making the Most of Human and Social Capital'. *Building Safer Cities: The Future of Disaster Risk, Disaster Risk Management Series* 13 (1 January 2003): 181–96.

Wisner, Ben. 'Five Years Beyond Sendai – Can We Get Beyond Frameworks?' *International Journal of Disaster Risk Science* 11, no. 2 (1 April 2020): 239–49. https://doi.org/10.1007/s13753-020-00263-0.

Wisner, Benjamin, J. C. Gaillard, Ilan Kelman and Ben Wisner, eds. *The Routledge Handbook of Hazards and Disaster Risk Reduction*. 1. publ. Routledge Handbooks. London: Routledge, 2012.

Worboys, Katherine. 'The Uses of History in Disaster Preparedness: The 1755 Lisbon Earthquake and the Construction of Historical Memory'. *Journal of the International Institute* 13, no. 2 (2006). http://hdl.handle.net/2027/spo.4750978.0013.203.

Zalik, Anna, and Isaac `Asume' Osuoka. 'Beyond Transparency: A Consideration of Extraction's Full Costs'. *The Extractive Industries and Society* 7, no. 3 (July 2020): 781–5. https://doi.org/10.1016/j.exis.2020.07.015.

Zulu, Eliya M., Donatien Beguy, Alex C. Ezeh, Philippe Bocquier, Nyovani J. Madise, John Cleland and Jane Falkingham. 'Overview of Migration, Poverty and Health Dynamics in Nairobi City's Slum Settlements'. *Journal of Urban Health: Bulletin of the New York Academy of Medicine* 88, no. Suppl 2 (June 2011): 185–99. https://doi.org/10.1007/s11524-011-9595-0.

INDEX